U0084848

晚餐 與 便當
一次搞定

1次煮2餐的
日式常備菜

作者－古靄茵

晚餐 與 便當
一次搞定

> 1次煮2餐的
> 日式常備菜

作　　者　古靄茵

攝　　影　幸浩生

編　　輯　鄭婷尹

美術設計　閻 虹

發 行 人　程安琪

總 策 畫　程顯灝

總 編 輯　呂增娣

主　　編　李瓊絲、鍾若琦

編　　輯　鄭婷尹、陳思穎、邱昌昊

美術總監　潘大智

美　　編　侯心苹、閻 虹

行銷總監　呂增慧

行銷企劃　謝儀方、吳孟蓉

發 行 部　侯莉莉

財 務 部　許麗娟

印　　務　許丁財

出 版 者　橘子文化事業有限公司

總 代 理　三友圖書有限公司

地　　址　106台北市安和路二段213號4樓

電　　話　(02) 2377-4155

傳　　真　(02) 2377-4355

E－mail　service@sanyau.com.tw

郵政劃撥　05844889 三友圖書有限公司

總 經 銷　大和書報圖書股份有限公司

地　　址　新北市新莊區五工五路2號

電　　話　(02) 8990-2588

傳　　真　(02) 2299-7900

製版印刷　皇城廣告印刷事業股份有限公司

初　　版　2015年11月

定　　價　新臺幣390元

I S B N　978-986-364-075-2(平裝)

本書繁體字版由香港萬里機構‧飲食天地出版社
授權在台灣地區出版發行。

◎版權所有‧翻印必究
　書若有破損缺頁 請寄回本社更換

國家圖書館出版品預行編目(CIP)資料

晚餐與便當一次搞定:1次煮2餐的日式常備菜 / 古靄茵
著. -- 初版. -- 臺北市：橘子文化, 2015.11
　面；　公分
ISBN 978-986-364-075-2(平裝)

1.食譜 2.日本

427.131　　　　　　　　　　　　　　104021784

家的味道是最好的味道

　　巨蟹座的女生，在傳統的潮州家庭長大，會是怎麼的一回事？從小跟外婆一起長大，被照顧得無微不至，一日三餐，吃著外婆的料理，「家的味道是最好的味道」根深柢固植在我的心裡。十年前，跟日本丈夫組織自己的小家庭，立志要把一直以來的信念帶到自己的小家庭內，要讓丈夫或日後的兒女知道，人生最幸福的事，就是吃著充滿愛的家庭料理。

　　初結婚時，我還是一整天在辦公室奔馳的女生，好不容易下班了，走到菜市場或超市買菜，再匆匆趕回家做飯，那時候只會做中菜，把菜買回家後，要醃要洗要煮，中菜的煮食過程要更繁複，同一菜式的食物先要分開處理，先蒸後炸，再加上菜湯，大炒一番。工作了一整天，拖著疲憊的身體弄一頓晚餐實在不是一件簡單的事。

　　後來認識了日本家庭料理，才知道處理日式家常菜舒服得多。一個小平底鍋，數個步驟，就已經做好一頓溫暖的晚飯。

　　謝謝萬里機構給予我這個機會，讓我能夠把自己的經驗跟大家分享。這本書記錄了我家飯桌的每一道家常菜，也收錄了我給丈夫做的便當。書裡的食譜特別為在職的你而設，兩口子的分量，簡樸的食材，淺易的步驟，希望大家能夠在繁忙的工作後，覺得筋疲力盡之時，還能夠以簡單而溫暖的家常味道慰藉疲憊的心靈。

　　也謝謝一直在身邊支持的每一位，因為有很多人的幫忙，這本食譜書才得以完成。感謝經常提點我的編輯小姐、專業的攝影大師哥哥、還有從不吝嗇地鼓勵我的朋友們，因為你們，我才能夠完成人生的第一本食譜書。

　　還要感謝我的外婆，感謝你把我培養成一個挑吃的女生；因為挑吃，所以才努力鑽研廚藝，令我在廚房內建立少少的成就。

　　感謝大家。

Candace

調味料百科

日本人相信食物有自己的味道，簡單的點綴就能夠做出美味的料理。
所以日本的調味料種類不多，以下介紹本書最常用的幾種。

日式醬油

分為濃口醬油及薄口醬油兩種，最常用的是濃口醬油，
但現代人較注重健康，所以也出現了減鹽醬油。本書食
譜皆使用濃口醬油。

味醂

帶甜味的料理酒。將一般料理酒內的甜度加倍，再加以
發酵而成。味醂不僅能夠提升料理的味道，還能夠軟化
肉類，是和風料理不可缺少的調味料。

料理酒

作料理用途的米酒，只需要少許分量就能夠發揮食物本
身的風味。日本人在宴客時會以上等的清酒作烹調用。

七味粉

正名為七味唐辛子，由 7 種香料混合而成，帶辣之餘又
不失香味。為日本最常見的其中一款餐桌調味料。

鰹魚粉

由乾鰹魚提煉而來，多用在醬汁或湯內，令味道更濃郁。
鰹魚粉水溶性很強，不需要特別處理。

鰹魚花（柴魚片）

跟鰹魚粉一樣，同樣由乾鰹魚提煉而來。但鰹魚花（柴魚片）呈木屑狀，味道較濃郁，多用作烹煮高湯。

味噌

味噌湯的基本味道來源。除了最常用的信州味噌外，還有赤味噌及白味噌兩種。本書使用最普遍的信州味噌。

醋

由米煉成的醋，沒有果醋的甜味，卻帶有清新的淡酸味。分有米醋及穀物醋兩種，米醋的味道較細緻，比穀物醋昂貴。

日式咖哩塊

日本的咖哩辣度不高，反而帶有甜味。咖哩塊分有不同的辣度，可以隨自己的口味選擇。

拌飯香鬆

日本的傳統食物，將小魚或海苔等食材壓碎，加入調味，最後再烘乾而成。拌飯香鬆多呈粉狀或粒狀，撒在米飯上一起食用，故稱為「拌飯香鬆」。拌飯香鬆備有很多不同口味，最普遍的為鰹魚味、明太子味及海苔雞蛋味。

便當盒的選擇

有精緻美味的食物，當然也要預備一個漂亮的便當盒。市場上的便當盒琳瑯滿目，令人目不暇給，不單只是材質選擇多，就連功能也日新月異，全面照顧不同的需求。除了材質的不同外，便當盒還分有單層、雙層甚至多層的設計，盒內的間隔也有不同，可以因應個人的需求及喜愛去選擇。

其實要選擇適合自己的便當盒一點都不困難，先弄清楚自己帶午飯便當的習慣及需要，再對照各類型便當盒的特性就可以了！

① 不鏽鋼便當盒

不鏽鋼便當盒的優點很多，它們堅固耐用、耐高溫、清洗容易，又不會殘留食物的氣味，無論放熱食或冷食都很合適。但不鏽鋼的便當盒只適合以蒸的方式或放入電鍋內加熱，不能夠以微波爐加熱食物。

② 木製／竹製便當盒

日本傳統的便當盒都以木或竹製成，時至今日，仍然很受日本人的歡迎。木或竹製的便當盒，清洗容易，還能夠透過吸收食物的油分，使便當盒更耐用。木製及竹製的便當盒可以用上數十年，很多老一輩的日本人，他們的木製或竹製的便當盒都是自小學時代已使用的了。但這類便當盒不能夠加熱，不適合愛吃熱騰食物的人。

③ 塑膠便當盒

現代人最常用的便當盒，無論圖案或款式都有很多。塑膠便當盒不僅輕便容易攜帶，還很耐熱，能夠放到微波爐加熱，是上班族的最貼心選擇。但一般的塑膠便當盒不容易清洗，還會殘留食物的氣味。坊間也有推出一種防油的塑膠便當盒，清洗比以前更方便，但售價也相對較貴。

• Bento&Co 京都本店

地址：604-8072 京都府京都市中京區六
　　　角通麩屋町東入八百屋町 117
網站：en.bentoandco.com

*以上未涉及商業行為，僅供參考。

· 本書使用的便當盒 → Bento&Co
Bento&Co 的便當盒款式多不勝數，有新穎有傳統，除了在日本京都的本店外，還設有網路商店，銷售到海外很多不同的地區。

和風的基本美味・日式高湯
Basic Japanese Broth

本書多道食譜都有用到日式高湯，但很多人為求方便，已把高湯簡化至將鰹魚粉入菜。其實日式高湯不難做，找一個週末，對自己對家人好一點，做一頓徹頭徹尾的美味和食，體驗一下日式高湯的不同之處。

材料
乾昆布 1 塊
鰹魚花（柴魚片）1/2 碗
水 5 碗

Ingredients
Kelp（Kombu）1pc
Bonito Flakes（Katsuo Bushi）1/2 bowl
Water 5 bowls

作法
1. 大鍋內加入水及乾昆布，浸泡 1 小時，再以小火煮熱。
2. 待水快要煮滾前將昆布取出，再轉大火至水大滾，關火。
3. 放入鰹魚花浸泡 5 分鐘，再過濾鰹魚花，留高湯使用。

Procedures
1.Soak kelp in pot with water for one hour, place the pot over medium low heat.
2.Remove kelp before water starts boiling. Increase to high heat till water boiled. Turn off.
3.Soak bonito flakes for 5 mins, strain the flakes through sieve.

 料理小提示 Cooking Tips

乾昆布及鰹魚花（柴魚片）在大型的日式超市均有販售。如果時間有限，又或高湯用量不多，還是建議以鰹魚粉加水代替自製高湯。
Kelp and Bonito Flakes can be found in Japanese supermarkets.Replace broth by dashi powder and water if you are running out of time.

我從來都不是大廚師。
不會做美侖美奐的菜，就只會做平平凡凡的家常菜。
用最簡單的調味，煮最日常的料理。

家常菜，就是讓家人吃了會感到幸福的料理。
沒有華麗的包裝，沒有名貴的味道；
只有做飯者的愛和心。

工作了一整天，拖著疲憊的身心，回到家中的飯桌；
與家人閒話家常，吃著那熟悉的味道；
手裡捧著一碗暖暖的湯，能夠把一切勞累融化。
這就是我的家常菜！

Chapter 1
雞肉料理

不同部位的雞肉各有適合的料理手法，經過
巧手烹調後，變成一道道美味上桌。

晚餐 親子丼

便當 唐揚炸雞

兩餐前準備 Preparation

材料
去骨雞腿 2 隻

醃料
日式醬油 1 湯匙
味醂 1 湯匙
料理酒 1 湯匙
糖 2 茶匙
薑汁 1 湯匙

Ingredients
Boneless Chicken Thigh 2 pcs

Marinade
Japanese Soy Sauce 1 tbsp
Mirin 1 tbsp
Sake 1 tbsp
Sugar 2 tsp
Ginger Sauce 1 tbsp

 作法 Procedures

雞肉切成一口大小，加醃料醃最少 20 分鐘。
Cut chicken thigh into bite size, marinade for at least 20 minutes.

🍚 親子丼
Oyakon Don
（*Chicken and Egg on Rice*）

🍲 唐揚炸雞 ———
Japanese Chicken Nuggets

 親子丼

Oyakon Don（*Chicken and Egg on Rice*）

材料

已準備的去骨雞腿肉 1 隻量
洋蔥 1 個
雞蛋 2 顆
白飯 2 碗
巴西里 少許

Ingredients

Prepared Chicken Thigh 1 pc
Onion 1 pc
Egg 2 pcs
Rice 2 bowls
Parsley little

調味料

日式醬油 2 湯匙
味酥 2 湯匙
糖 2 茶匙
水 300 毫升
鰹魚粉 2 茶匙

Seasoning

Japanese Soy Sauce 2 tbsp
Mirin 2 tbsp
Sugar 2 tsp
Water 300 ml
Dashi Powder 2 tsp

作法

1. 洋蔥切片成絲；打發雞蛋；調味料放小碗內拌勻。
2. 小鍋內放少許油，炒洋蔥絲至軟，放入雞肉大火炒至表面金黃。
3. 將作法 1 調好的調味料加入小鍋內，蓋上鍋蓋，小火煮約 10 分鐘。
4. 開蓋倒入蛋液，輕輕拌勻，再蓋上煮約 3 分鐘。
   （輕輕將蛋液推開便可，避免不停翻動蛋液。）
5. 將雞肉及雞蛋盛在白飯上，吃時可搭配巴西里。

Procedures

1. Slice onion. Beat eggs in a small bowl. Mix marinade in another small bowl.
2. Heat a small pot with oil, stir fry onion till tender. Add chicken, stir fry till brown.
3. Pour marinade into pot, cover and simmer for 10 minutes.
4. Pour in beaten egg, stir gently, cover and simmer for another 3 minutes.（Do not over stir the eggs.）
5. Place chicken and egg on rice after 2 mins. Serve with parsley.

 料理小提示 Cooking Tips

日本正宗的親子丼，蛋液分兩次倒入，讓雞蛋有半熟口感。但我們還是衛生健康一點，把雞蛋弄個全熟。
In Japan, the eggs for Oyako Don are served half-cooked. I would suggest to have the eggs well-cooked due to hygiene problem.

利用簡單的材料，在短時間內做出
來的美味料理。是繁忙的上班族不
可缺少的食譜。

很受歡迎的一道料理，除了適合作便當主食外，還可以當小吃，在派對或跟朋友聚會時品嘗最適合不過。

唐揚炸雞
Japanese Chicken Nuggets

材料
已準備的去骨雞腿肉 1 隻量
玉米粉 適量

Ingredients
Prepared Chicken Thigh 1 pc
Corn Starch little

作法
1. 小鍋內燒滾油。
2. 玉米粉放小碗內,先將雞塊均勻裹粉,再放入滾油中以大火炸至表面金黃,盛起。
3. 雞塊放涼後,再燒熱油,放入雞塊以小火慢炸至內裡熟透,再轉大火將雞塊的油分逼出。

Procedures

1.Heat oil in a small pot.

2.Coat the chicken pieces with corn starch. Deep fried in hot boiling oil.

3.After the chicken cooled, bring to deep fry over low heat again. Turn on high heat right before done.

料理小提示 Cooking Tips

吃的時候淋點新鮮檸檬汁,可以去油膩。
Drizzle with fresh lemon juice for less greasy texture.

晚餐 便當

雞肉漢堡

鮮冬菇釀雞肉

兩餐前準備 Preparation

材料
雞絞肉 300 克
雞蛋 1 顆
鮮奶 1 湯匙

Ingredients
Minced Chicken 300g
Egg 1 pc
Milk 1 tbsp

醃料
日式醬油 1 湯匙
味酥 1 湯匙
料理酒 1 湯匙
糖 2 茶匙
胡椒粉 1 茶匙
玉米粉 1 湯匙

Marinade
Japanese Soy Sauce 1 tbsp
Mirin 1 tbsp
Sake 1 tbsp
Sugar 2 tsp
Pepper 1 ts
Corn Starch 1 tbsp

作法 Procedures

大碗內，放雞絞肉、雞蛋、鮮奶及醃料拌勻。
Mix minced chicken, egg, milk and marinade in a big bowl.

雞肉漢堡 *Chicken Burger Steak*

鮮冬菇釀雞肉 ——
Mushroom Stuffed with Chicken

晚餐

雞肉漢堡
Chicken Burger Steak

材料
已準備的雞絞肉 150 克
紅蘿蔔 1/2 條
葱 1 株

調味料
日式醬油 1 湯匙
味醂 1 湯匙
糖 1 茶匙

Ingredients
Prepared Minced Chicken 150g
Carrot 1/2 pcs
Green Onion 1 stalk

Seasoning
Japanese Soy Sauce 1 tbsp
Mirin 1 tbsp
Sugar 1 tsp

作法
1. 紅蘿蔔刨絲,葱切小粒。
2. 紅蘿蔔及葱粒加至雞絞肉內拌勻。
3. 將肉分為 4 等份,用手將其中一份搓成圓餅狀,重複做出其餘 3 份。
   (把手弄濕一點再搓肉,肉餅會更整齊美觀。)
4. 將肉餅放到已燒熱的平底鍋上,用大火煎封上下兩面,再加 2 湯匙水,蓋上蓋,轉小火煮約 5 分鐘,盛起。
5. 調味料放小碗內拌勻,吃時加在雞肉漢堡上。

Procedures
1.Shred carrot. Finely dice green onion.
2.Mix minced chicken with carrot and green onion.
3.Divide meat paste into 4 portions. Roll meat paste into shape of round cake.(Wet hands with some water when shaping meat cakes.)
4.Pan fry meat cake over high heat till both sides turn golden brown. Add 2 tbsp of water, cover and reduce to low heat. Cook for 5 mins.
5.Mix seasoning in small bowl, drizzle over chicken burger steak when serve.

 料理小提示 Cooking Tips

想增添多一點和味,可以在已完成的雞肉漢堡上放一葉紫蘇,再放一點點蘿蔔泥,淋上調味好的醬汁一起品嘗。
Place a piece of Japanese basil and shredded radish on burger steak for more Japanese style and taste.

這道菜是牛肉漢堡排的變奏，感覺更輕盈，味道更清新。

冬菇跟蝦米的香味混在一起，
很有中式家常菜味道的料理！

22

鮮冬菇釀雞肉

Mushroom Stuffed with Chicken

材料
已準備的雞絞肉 150 克
鮮冬菇 6 朵
蝦米 1 湯匙
葱 1 株

Ingredients
Prepare Minced Chicken 150g
Fresh Shitake Mushroom 6 pcs
Dried Shrimp 1 tbsp
Green Onion 1 stalk

調味料
日式醬油 1/2 湯匙
味醂 1 湯匙
糖 1 茶匙
水 50 毫升
鰹魚粉 1 茶匙

Seasoning
Japanese Soy Sauce 1/2 tbsp
Mirin 1 tbsp
Sugar 1 tsp
Water 50ml
Dashi Powder 1 tsp

作法
1. 鮮冬菇洗淨去蒂頭，用刀在菇面切出十字。
   （劃了十字的冬菇較可愛，若怕麻煩則可省去。）
2. 蝦米略泡水浸軟後切碎，葱切小粒，調味料放小碗內拌勻。
3. 蝦米及葱加入雞絞肉內拌勻。
4. 將雞肉均勻地釀入鮮冬菇內。
5. 平底鍋加少許油燒熱，將雞肉面向下，煎至雞肉呈金黃色。
6. 翻面稍煎鮮冬菇的一面，加入作法 2 調好的調味料，蓋上鍋蓋，轉小火，烹煮至湯汁濃稠。

Procedures
1. Rinse shitake mushroom, cut a cross on top of every shitake mushroom.
   （The cross gives a better presentation to shitake mushroom.）
2. Soak dried shrimp in water, drain and finely dice. Finely chop green onion. Mix seasoning in a small bowl.
3. Mix chicken with shitake mushroom and dried shrimp.
4. Stuff shitake mushroom with meat paste.
5. Heat frying pan with oil, cook stuffed shitake mushroom with the side of meat paste face down till golden brown.
6. Turn over shitake mushroom, add seasoning. Cover and reduce to low heat, cook till sauce thicken.

 料理小提示 Cooking Tips

這道料理跟日式高湯煮的蔬菜最搭配。
The dish is perfectly match with broth boiled vegetable.

晚餐 番茄雞肉煮

便當 雞肉雜菜卷

兩餐前準備 Preparation

材料
去骨雞腿排 4 塊

Ingredients
Boneless Chicken Thigh 4 pcs

醃料
鹽 2 茶匙
黑胡椒粉 2 茶匙

Marinade
Salt 2 tsp
Black Pepper 2 tsp

 作法 Procedures

將整塊去骨雞腿排去皮及脂肪，加醃料醃至少 20 分鐘。
Remove skin and fat from chicken thigh, marinate for at least 20 minutes.

番茄雞肉煮
Braised Chicken with Tomato

 雞肉雜菜卷
*Chicken Roll with
Assorted Vegetables*

晚餐

番茄雞肉煮
Braised Chicken with Tomato

材料
已準備的去皮雞腿排 2 塊
番茄 3 個
洋蔥 1/2 個
蒜蓉 1 茶匙
玉米粉 2 茶匙
巴西里 少許

調味料
糖 2 茶匙
鹽 1 茶匙
胡椒粉 1 茶匙

Ingredients
Prepared Chicken Thigh 2 pcs
Tomato 3 pcs
Onion 1/2 pcs
Minced Garlic 1 tsp
Corn Starch 2 tsp
Parsley little

Seasoning
Sugar 2 tsp
Salt 1 tsp
Pepper 1 tsp

作法
1. 小鍋內煮滾水，放入番茄煮約 3 分鐘，盛起後將番茄去皮，並切成小丁。
   （番茄煮久一點，之後就可以縮短炒番茄的時間。）
2. 雞排切成一口大小，洋蔥切片成絲，巴西里切碎，玉米粉加水 1 湯匙拌勻。
3. 平底鍋內加油燒熱，爆香蒜蓉，再放洋蔥炒至軟，加雞肉炒至八成熟。
4. 放入番茄及調味料，炒至番茄變泥，加作法 2 調好的玉米粉水勾芡。
5. 吃時搭配巴西里碎。

Procedures
1. Boil tomato in a small pot of water for 3 minutes. Remove skin from tomato, dice. (The longer the tomato cooking time, the shorter the stir fry tomato time in next step.)
2. Cut chicken into bite size. Slice onion. Finely chop parsley. Mix corn starch with 1 tbsp of water.
3. Heat frying pan with oil, saute garlic and onion till tender. Add chicken, stir fry.
4. Add tomato and seasoning, stir fry till tomato tender. Add corn starch mix, stir fry till sauce thicken.
5. Drizzle with parsley.

 料理小提示 Cooking Tips

脂肪熱量很低的一道料理，多吃也無妨。想來點新意，可以搭配義大利麵或法式麵包，和一些氣泡礦泉水，很健康的一道晚餐呢！
A very low calories dish. Best with spaghetti or bread, with a bottle of sparkling water.

酸酸甜甜的一道菜，屬於日本的西洋
料理，很適合炎夏食用呢！

看似困難，其實極其容易的一道料理，又綠又紅的肉卷，放在便當盒內很搶眼呀！

雞肉雜菜卷

Chicken Roll with Assorted Vegetables

材料
已準備的去皮雞腿排 2 塊
紅蘿蔔 1 條
四季豆 10 條

蔬菜煮料
日式醬油 1 湯匙
味酥 1 湯匙
水 200 毫升
鰹魚粉 1 茶匙

Ingredients
Prepared Chicken Thigh 2 pcs
Carrot 1 pc
Green Beans 10 pcs

Seasoning for vegetables
Japanese Soy Sauce 1 tbsp
Mirin 1 tbsp
Water 200 ml
Dashi Powder 1 tsp

調味料
日式醬油 1 湯匙
味酥 2 湯匙
料理酒 2 湯匙
糖 2 茶匙

Seasoning
Japanese Soy Sauce 1 tbsp
Mirin 2 tbsp
Sake 2 tbsp
Sugar 2 tsp

作法
1. 紅蘿蔔去皮，切去頭尾，再切成約 5 公釐厚的粗條狀。
2. 小鍋內加入蔬菜煮料，水滾後放紅蘿蔔條煮約 10 分鐘，四季豆略煮至軟。
3. 雞肉打直放砧板上，用刀在較厚的位置切數下，可以讓雞排更快熟。
4. 將紅蘿蔔條及四季豆放在雞排的末端，再向上捲。用竹籤固定開口。記得要捲得緊一點，完成後的雞肉卷才會漂亮。
5. 平底鍋加油燒熱，將雞肉卷的開口朝下入鍋，大火煎熟，再翻面煎其餘部位至金黃。
6. 調味料放小碗內攪勻，加入平底鍋內，煎至收汁。

Procedures
1.Peel carrot, cut into 5mm wide sticks.
2.Boil seasoning for vegetables in a small pot, add carrots cook for 10 minutes, and greens beans for a while.
3.Add few cuts on the thick parts of chicken thigh. It will speed up the cooking time of chicken.
4.Roll carrot sticks and green beans with chicken thigh. Use a skewer to seal the open end.
5.Heat frying pan with oil, place the open end of the chicken down, cook over high heat. Roll and cook the whole chicken roll.
6.Stir in seasoning, cook till sauce thicken.

 料理小提示 Cooking Tips
雞肉卷切片後再放進盒內，讓美麗的顏色呈現出來。
Slice the chicken roll before put inside lunchbox.

晚餐　便當

京葱
雞肉串燒

照燒雞排

兩餐前準備 Preparation

材料
去骨雞腿排 3 塊

醃料
日式醬油 2 湯匙
味醂 1 湯匙

Ingredients
Boneless Chicken Thigh 3 pcs

Marinade
Japanese Soy Sauce 2 tbsp
Mirin 1 tbsp

 作法 Procedures

整塊雞腿排，不用切小，醃至少 20 分鐘。
Marinade the whole chicken thigh for at least 20 minutes.

京葱雞肉串燒 *Leek and Chicken Skewers*

照燒雞排 *Teriyaki Chicken*

京葱雞肉串燒

Leek and Chicken Skewers

材料
已準備的去骨雞腿排 1 塊
京葱（大葱）1 株

Ingredients
Prepared Chicken Thigh 1 pc
Leek 1 pc

調味料
日式醬油 1 湯匙
味醂 1 湯匙
料理酒 1 湯匙
糖 2 茶匙

Seasoning
Japanese Soy Sauce 1 tbsp
Mirin 1 tbsp
Sake 1 tbsp
Sugar 2 tsp

作法
1. 雞肉切成一口大小，京葱也切成跟雞肉一樣的長度。
2. 調味料放小碗內拌勻，備用。
3. 利用竹籤相間地串起雞肉及京葱。（竹籤不要太長，不能比平底鍋長。）
4. 平底鍋不用加油，燒至稍熱，放入京葱雞肉串，先以大火煎封雞肉兩面。
5. 轉小火慢煎，期間不停均勻地刷上作法 2 的調味料。
6. 至雞肉全熟後，再轉大火煎至收汁。

Procedures
1.Cut chicken into bite size. Cut leek into same size with chicken
2.Mix seasoning in a small bowl.
3.Skewer chicken and leek.
   （The skewer must not be longer than the frying pan.）
4.Heat frying pan without oil, add skewer. Cook over high heat till both sides golden brown.
5.Reduce to low heat, brush seasoning on skewers.
6.Turn to high heat when chicken well cooked.

 料理小提示 Cooking Tips

多做幾款串燒，把家裡變成最舒適、最溫馨的居酒屋。
Make different kind of skewers, turn your kitchen to a homey izakaya.

有時候，兩口子工作都倦了，回到家裡，吃著
自家做的和式串燒，喝一口冰涼啤酒，輕輕鬆
鬆，精神都回來了。

人見人愛的一道料理，是
和式家常菜的基本，要做
得好，一點都不困難呢！

照燒雞排
Teriyaki Chicken

材料
已準備的去骨雞腿排 2 塊

Ingredients
Prepared Chicken Thigh 2 pcs

調味料
日式醬油 1 湯匙
味醂 1 湯匙
料理酒 1 湯匙
糖 2 茶匙

Seasoning
Japanese Soy Sauce 1 tbsp
Mirin 1 tbsp
Sake 1 tbsp
Sugar 2 tsp

作法
1. 調味料放小碗內拌勻，備用。
2. 平底鍋不用加油，稍微燒熱，放入雞腿排，雞皮向下，先以大火煎封雞肉兩面。
（雞皮有很多油脂，所以不用額外加油煎。）
3. 轉小火慢煎，期間不停均勻地刷上作法 1 的調味料。
4. 至雞肉全熟後，再轉大火煎至收汁。
5. 取出放涼，再切成小塊。

Procedures
1. Mix seasoning in a small bowl.
2. Heat frying pan without oil, add chicken thigh with the skin face down. Cook over high heat till both sides golden brown.（No extra oil is needed when cooking chicken with skin, as it contains lots of oil and fat.）
3. Reduce to low heat, brush seasoning on chicken thigh.
4. Turn to high heat till sauce thicken.
5. Cut chicken thigh into pieces after cooled down.

料理小提示 Cooking Tips

烹調後的平底鍋先不要清洗，切一個小青椒，慢慢煎煮，讓雞腿排的精華及醬汁都沾到青椒上，放入便當盒一起當作午飯。
Do not waste the yummy sauce in the pan. Slice a bell pepper and cook in the same pan of teriyaki chicken, let the bell pepper absorb the remaining sauce. Place inside lunchbox together with chicken.

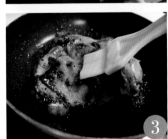

晚餐　便當

吉列南瓜
雞肉餅

南瓜雞腿
甘酢煮

兩餐前準備 Preparation

材料
雞腿 10 隻
去骨雞腿肉 1 塊

Ingredients
Chicken Drumsticks 10 pcs
Boneless Chicken Thigh 1 pc

 作法 Procedures

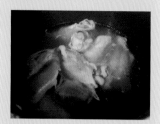

1. 去骨雞腿肉去皮。
2. 小鍋內燒滾水，放入雞腿及去骨雞腿肉汆燙，盛起。
1. Remove skin from chicken thigh
2. Rinse chicken drumsticks and chicken thigh, scald in boiling water for a
   short while. Rinse again. Set aside.

 料理小提示 Cooking Tips

這兩道菜的靈魂在於湯汁，建議直接製作日式高湯使用，作法可參考 p.9。
Since this menu consists of large amount of broth, suggested to make broth
rather than instant dashi powder. Please refer to page 9.

南瓜雞腿甘酢煮 *Braised Chicken Drumsticks with Pumpkin*

吉列南瓜雞肉餅—
Chicken and Pumpkin Croquet

南瓜雞腿甘酢煮

Braised Chicken Drumsticks with Pumpkin

材料
已準備的雞腿 10 隻
日本南瓜 1 個

Ingredients
Prepared Chicken Drumsticks 10 pcs
Japanese Pumpkin 1 pc

調味料
日式醬油 3 湯匙
味醂 1 湯匙
料理酒 1 湯匙
冰糖 5 ～ 6 粒
薑 2 片
日式高湯 500 毫升
（或水 500 毫升＋鰹魚粉 2 茶匙）

Seasoning
Japanese Soy Sauce 3 tbsp
Mirin 1 tbsp
Sake 1 tbsp
Rock sugar 5-6 pcs
Ginger 2 slices
Japanese Broth 500ml
（It can be replaced by 500ml water ＋ 2 tsp of dashi powder.）

作法
1. 南瓜洗淨去籽，切成大約 2.5 公分╳2.5 公分大小。
2. 小鍋內，加入調味料煮滾。
3. 放入雞腿及南瓜，小火烹煮約 20 分鐘。
　（記著要把便當的去骨雞腿肉一併放進去煮。）

Procedures
1.Remove seed from pumpkin, cut into 1"x 1" pieces.
2.Boil seasoning in a pot.
3.Add chicken drumsticks and pumpkin, braise for 20 minutes.
（Remember to add chicken thigh together in the same time!）

 料理小提示 Cooking Tips

南瓜分量不少，享受美食之餘，記得要把半個南瓜和去骨雞腿肉盛起放涼，做明天的便當呀！
This is a big portion of pumpkin, Reserve chicken thigh and half portion of pumpkin for lunchbox cooking.

非常省時的一道料理，把晚餐和明天便當的
材料一併放進大鍋裡煮，方便又美味！

這個前一天晚上做就可以
了，功夫少又可口，謹記不
要嘴饞把料理當宵夜吃掉！

吉列南瓜雞肉餅

Chicken and Pumpkin Croquet

材料
「南瓜雞腿甘酢煮」剩餘的
去骨雞腿肉及南瓜
玉米粉 2 湯匙
雞蛋 1 顆
麵包粉 1 碗

調味料
鹽 1/2 茶匙
味醂 1/2 湯匙

Ingredients
Chicken Thigh and Pumpkin
from previous menu
Corn Starch 2 tbsp
Egg 1 pc
Bread Crumb 1 bowl

Seasoning
Salt 1/2 tbsp
Mirin 1/2 tbsp

作法
1. 南瓜放大碗內用叉子壓碎，雞腿肉用手撕成絲，跟南瓜及調味料拌勻。
2. 用手將南瓜雞肉泥分為 4 等份，用手搓成圓餅狀。
3. 依次序將南瓜雞肉餅沾上玉米粉、蛋液及麵包粉。
4. 小鍋內燒滾油，放南瓜雞肉餅炸至金黃即可。
　 （因南瓜及雞肉都已熟透，所以只需稍稍炸至外表金黃。）

Procedures
1. Mash pumpkin, shred chicken. Mix pumpkin, chicken and seasoning.
2. Divide meat paste into 4 equal portions. Roll meat paste into shape of oval cake.
3. Coat meat cake with corn starch, beaten egg and bread crumbs.
4. Deep fried meat cakes in boiling oil till golden brown.
　（It does not need a long cooking time as pumpkin and chicken are already cooked.）

 料理小提示 Cooking Tips

可以前一天預先做，第二天早上用烤爐稍烤一下，又或炸完的油不要
倒掉，早上再翻炸至熱便可。
Croquet can be made the day before. Reheat by oven or deep fry again
in the morning.

新相識的朋友總愛問：「你在家都做日本菜嗎？」
我點點頭。

初結婚時，我其實在做中國菜，做跟你家或我家一樣的家常菜。
每天晚上，和丈夫如常地吃著中式的晚飯。
有一天，不知哪來的念頭，給他做了一頓粗糙的日式晚餐；
他大口大口吃下肚，沒有多說話，我卻把一切看在眼裡。

一個人離開家鄉，離開自己的家庭料理，
每天在陌生的國度裡，吃著陌生的食物。
「有感到孤單嗎？」一個永遠只放在我心裡的問題。

由那時候開始，我努力學習各式各樣日式家常菜。
讓這個一直為我努力工作的人，
每天回到家裡，都能夠嘗到親切的味道。

Chapter 2
豬肉料理

用鮮甜豬肉製成的豬排、肉丸、滷肉……等菜肴，餵飽下班後的疲累精神，恢復滿滿元氣。

晚餐 便當

吉列豬排

薑汁豬肉燒

兩餐前準備 Preparation

材料	Ingredients
嫩肩里肌肉排 4 塊	Boneless Pork Chop 4 pcs

醃料	Marinade
日式醬油 3 湯匙	Japanese Soy Sauce 3 tbsp
味醂 1 湯匙	Mirin 1 tbsp
料理酒 1 湯匙	Sake 1 tbsp
糖 2 茶匙	Sugar 2 tsp

作法 Procedures

1. 嫩肩里肌肉排利用刀背剁鬆。
2. 放大碗內，加醃料醃至少 20 分鐘。

1. Use edge of knife to massage pork chops until tender.
2. Mix pork chops with marinade, set aside for at least 20 minutes.

吉列豬排 *Deep Fried Breaded Pork Cutlets*

薑汁豬肉燒
Pork Saute in Ginger

晚餐

吉列豬排
Deep Dried Breaded Pork Cutlets

材料
已準備的嫩肩里肌肉排 2 塊
玉米粉 2 湯匙
雞蛋 1 顆
麵包粉 1 碗

Ingredients
Prepared Pork Chop 2 pcs
Corn Starch 2 tbsp
Egg 1 pc
Bread Crumbs 1 bowl

作法
1. 小碗內打發雞蛋。大鍋裡加油燒熱。
2. 依次序將嫩肩里肌肉排沾上玉米粉、蛋液及麵包粉。
3. 放滾油內以中小火慢炸，待變得熟透後，轉大火炸至金黃。
   最後再以大火將肉排內的油分逼出。

Procedures
1.Beat egg in a small bowl. Add oil in a pot, bring to boil.
2.Coat pork chop with corn starch, beaten egg and bread crumbs.
3.Put pork chops into hot oil and deep fry over low heat till golden.
   Turn on high heat before dish up.（High heat right before dish up helps
   releasing oil from pork chop.）

 料理小提示 Cooking Tips

不要忘記切得又細又薄的高麗菜絲沙拉，跟吉列豬排很搭配呢！
Finely sliced lettuce salad always the best match with Deep Fried Breaded Pork Cutlets.

想做到跟餐廳一樣香脆軟嫩的吉列豬排，祕訣
在於先把豬排完全拍鬆和烹調時火候的控制。

感覺很男性化的一道料理，每次我
都把丈夫的便當盒裝得滿滿的。

薑汁豬肉燒
便當

Pork Saute in Ginger

材料
已準備的嫩肩里肌肉排 2 塊
洋葱 1/2 個
青椒 1/2 個

Ingredients
Prepared Pork Chop 2 pcs
Onion 1/2 pcs
Bell Pepper 1/2 pcs

調味料
日式醬油 2 湯匙
料理酒 2 湯匙
薑汁 1 湯匙

Seasoning
Japanese Soy Sauce 2 tbsp
Sake 2 tbsp
Ginger Sauce 1 tbsp

作法
1. 洋葱及青椒切絲。
2. 平底鍋放油燒熱，加入洋葱及青椒炒軟，盛起備用。
3. 在同個平底鍋內，放嫩肩里肌肉排煎至八分熟，將作法 2 的炒洋葱及青椒回鍋。
4. 下調味料，煎至豬排全熟及收汁。

Procedures
1. Slice onion and bell pepper.
2. Heat oil in frying pan, cook onion and green pepper till tender, set aside.
3. Pan fry pork chops in same frying pan. Add onion and green pepper when pork chops are 80% done.
4. Add seasoning, cook till pork chops well done and sauce thicken.

  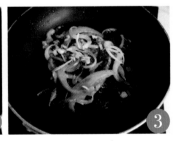

 料理小提示 Cooking Tips

若喜歡薑的香味，可自行加重薑汁的分量。
Drizzle with fresh lemon juice for less greasy texture.

晚餐　便當

汁燒滷肉　和風泡菜滷肉湯

兩餐前準備 Preparation

材料
五花肉 3 條

醬料
日式高湯 700 毫升
料理酒 20 毫升
薑 4 片
蔥 2 株
八角 3 顆

Ingredients
Pork Belly 3 pcs（around 2" in width）

Sauce
Japanese Broth 700ml
Sake 20ml
Ginger 4 slices
Green Onion 2 stalks
Star Anise 3 pcs

 作法 Procedures

1. 五花肉汆燙。
2. 小鍋內加醬料煮滾，放入五花肉，煮滾後轉小火，續煮 1.5 小時。

1. Rinse pork belly and scald in boiling water for a short while. Rinse again. Set aside.
2. Bring seasoning to boil, add pork belly. Braise for 1.5 hours.

 和風泡菜滷肉湯
Braised Pork Belly and
Kimchi in soup

汁燒滷肉 ——
Braised Pork Teriyaki

晚餐

和風泡菜滷肉湯
Braised Pork Belly and Kimchi in soup

材料
已準備的五花肉 1 條
韭菜 150 克
黃豆芽 100 克
韓國泡菜 100 克
白芝麻 1 茶匙

Ingredients
Prepared Pork Belly 1 pc
Chive 150g
Bean Sprouts 100g
Korean Kimchi 100g
Sesame 1 tsp

調味料
日式醬油 2 湯匙
糖 2 茶匙
胡椒粉 1 茶匙
日式高湯 500 毫升

Seasoning
Japanese Soy Sauce 2 tbsp
Sugar 2 tsp
Pepper 1 tsp
Japanese Broth 500ml

作法
1. 五花肉切厚片，韭菜切成 5 公分長。
2. 小鍋內放油燒熱，加五花肉略炒，再放入泡菜及黃豆芽炒勻，炒至黃豆芽變軟。
3. 加入調味料及韭菜，蓋上，煮至滾。
4. 盛盤，吃時撒上白芝麻。

Procedures
1.Thickly slice pork belly. Cut chive into 5cm long.
2.Heat oil in a pot, stir fry pork belly. Add kimchi and bean sprouts, cook till bean sprouts tender.
3.Add seasoning and chive, cover and cook till boiling.
4.Dish up, drizzle with sesame when serve.

 料理小提示 Cooking Tips
黃豆芽在一般超市或市場較少見，可以用其他豆芽菜代替！
Every kind of bean sprouts can be used in this dish.

日式的鹹香味，跟韓式的辛辣味，互相搭配，帶出不一樣的驚喜味道。

很省工夫的一道主食，賴床還可以趕得
及完成的一個便當料理！

54

汁燒滷肉
便當

Braised Pork Teriyaki

材料
已準備的五花肉 1 條
秀珍菇 1 盒
蒜蓉 1 茶匙

調味料
日式醬油 2 湯匙
味醂 2 湯匙
料理酒 2 湯匙
水 2 湯匙

Ingredients
Prepared Pork Belly 1 pc
Fresh Shimeji Mushroom 1 pack
Minced Garlic 1 tsp

Seasoning
Japanese Soy Sauce 2 tbsp
Mirin 2 tbsp
Sake 2 tbsp
Water 2 tbsp

作法
1. 五花肉切薄片，秀珍菇去蒂頭擦乾淨。
   （菇類不要用水沖洗，用廚房紙巾擦乾淨便可。）
2. 平底鍋加油燒熱，爆香蒜蓉。
3. 下秀珍菇炒至軟。
4. 加入五花肉及調味料，炒至收汁。

Procedures
1.Thinly slice pork belly. Remove root from shimeji mushroom.
   （Do not rinse fresh mushroom, wipe with kitchen paper only.）
2.Heat oil in frying pan, saute garlic.
3.Stir fry shimeji mushroom still tender.
4.Add pork belly and seasoning, stir fry till sauce thicken.

 料理小提示 Cooking Tips

這道料理的搭配很隨意，也可以將秀珍菇換成其他菇類或蔬菜。
Change recipe to other choice of mushroom or vegetables according to your own preference.

晚餐 便當

咖哩吉列雞蛋肉卷

和風豬肉蒸蛋白

材料
豬絞肉 300 克

Ingredients
Minced Pork 300g

醃料
日式醬油 1 湯匙
味醂 1 湯匙
糖 2 茶匙

Marinade
Japanese Soy Sauce 1 tbsp
Mirin 1 tbsp
Sugar 2 tsp

作法 Procedures

大碗內放豬絞肉及醃料拌勻。
Mix minced pork and marinade in a bowl.

56

└─ 🍲 和風豬肉蒸蛋白 *Steam Egg White with Pork*

└─ 🍲 咖哩吉列雞蛋肉卷
Deep Fried Breaded Curry Meat Ball with Egg

晚餐

和風豬肉蒸蛋白
Steam Egg White with Pork

材料
已準備的豬絞肉 100 克
嫩豆腐 1 塊
蛋白 3 顆
葱 1 株

調味料
日式醬油 1/2 湯匙
料理酒 1 湯匙
水 100 毫升
鰹魚粉 1 茶匙

Ingredients
Prepared Minced Pork 100g
Soft Tofu 1 pc
Egg White 3 pcs
Green Onion 1 stalk

Seasoning
Japanese Soy Sauce 1/2 tbsp
Sake 1 tbsp
Water 100ml
Dashi Powder 1 tsp

作法
1. 葱剁碎;調味料放小碗內拌勻。
2. 嫩豆腐整塊隔水蒸 10 分鐘,瀝去水分後切塊。
3. 蛋白加鹽拌勻,放盤內,加入豆腐塊,蓋上錫箔紙,隔水以小火蒸 15 ～ 20 分鐘。(用錫箔紙覆蓋豆腐再蒸,可以保持豆腐表面平滑。)
4. 平底鍋加油燒熱,放進豬絞肉炒至八分熟,加作法 1 拌好的調味料炒至肉全熟及收汁。
5. 將炒好的豬絞肉放在已蒸好的蛋白上,撒上葱花。

Procedures
1. Finely chop green onion. Mix seasoning in small bowl.
2. Steam soft tofu for 10 minutes, drain and dice.
3. Mix egg white with salt. Place in a dish, add tofu cover with foil. Steam for 15-20 minutes. (Cover with foil helps to keep the surface smooth.)
4. Heat oil in frying pan, stir fry minced pork until 80% done. Add seasoning and stir fry till sauce thicken.
5. Place cooked minced pork on steamed egg white. Drizzle with green onion.

 料理小提示 Cooking Tips

剩下的蛋黃一點都不浪費,可以用在下一個便當料理上呀!
The left-over egg yolks can be used in next dish.

清新的一道料理。蛋白的香滑，加上夠味的汁
燒絞肉，令人無法抗拒。

炸得香脆可口的麵包粉，包著香濃的咖哩肉，再加上幼滑的煮雞蛋，一顆夠嗎？

咖哩吉列雞蛋肉卷

Deep Fried Breaded Curry Meat Ball with Egg

材料
已準備的豬絞肉 200 克
雞蛋 2 顆
日式咖哩粉 2 茶匙
玉米粉 2 湯匙
麵包粉 1 碗
「和風豚肉蒸蛋白」剩下的蛋黃

Ingredients
Prepared Minced Pork 200g
Egg 2 pcs
Japanese Curry Powder 2 tsp
Corn Starch 2 tbsp
Bread Crumbs 1 bowl
Egg Yolks from previous dish

作法
1. 雞蛋放滾水內煮熟後剝殼。另外在小碗內打發蛋黃。
2. 豬絞肉放大碗內,加咖哩粉拌勻。
3. 將作法 2 的豬絞肉分為 2 等份,先用手壓扁其中一份,煮好的雞蛋放在中間,再將絞肉包裹整顆雞蛋;另一份重複同樣步驟完成。
4. 包好的雞蛋肉卷依序沾上玉米粉、蛋液及麵包粉。
5. 放入滾油炸至金黃。（先用小火慢炸,最後才轉大火,可保持肉卷美觀）

Procedures
1. Boil eggs and shell. Beat egg yolks in a bowl.
2. Mix minced pork with curry powder.
3. Divide minced pork into 2 equal portions. Flatten one portion by hand, place boiled egg in the middle, wrap by meat paste.
4. Coat the meat ball by corn starch, beaten egg yolks and bread crumbs.
5. Deep fry in boiling oil till golden brown.（Cook with low heat at first and turn to high heat before done.）

 料理小提示 Cooking Tips

雞蛋不要太大顆,若肉卷太大會很難放進便當盒內。
Try to cook with eggs of small size.

晚餐　　便當

咕咾肉丸子

高麗菜肉卷

兩餐前準備 Preparation

材料
豬絞肉 300 克
蔥 1 株
紅蘿蔔 1/2 條

醃料
日式醬油 2 湯匙
味醂 2 湯匙
糖 2 茶匙
玉米粉 2 湯匙

Ingredients
Minced Pork 300g
Green Onion 1 stalk
Carrot 1/2 pcs

Marinade
Japanese Soy Sauce 2 tbsp
Mirin 2 tbsp
Sugar 2 tsp
Corn Starch 2 tbsp

 作法 Procedures

1. 蔥剁碎，紅蘿蔔刨絲。
2. 大碗內放豬絞肉、蔥碎、紅蘿蔔絲及醃料，攪拌至帶有黏性。
1. Finely chop green onion. Shred carrot.
2. Mix minced pork, green onion, carrot and marinade.

高麗菜肉卷 *Pork Wrapped in Cabbage*

咕咾肉丸子
*Meat Ball in Sweet and
Sour Sauce*

高麗菜肉卷
Pork Wrapped in Cabbage

材料
已準備的豬絞肉 150 克
高麗菜 6 葉
紅蘿蔔 1/2 條
鴻喜菇 1 盒

調味料
日式醬油 1 湯匙
味醂 1 湯匙
日式高湯 200 毫升
鰹魚粉 1 茶匙

Ingredients
Prepared Minced Pork 150g
Cabbage 6 pcs
Carrot 1/2 pcs
Shimeji Mushroom 1 pack

Seasoning
Japanese Soy Sauce 1 tbsp
Mirin 1 tbsp
Japanese Broth 200ml
Dashi Powder 1 tsp

作法
1. 紅蘿蔔切厚片，利用壓模壓成花形；鴻喜菇去蒂頭並擦乾淨。
   （將紅蘿蔔切花是為了更美觀，也可省去。）
2. 小鍋內煮滾水，放高麗菜汆燙至軟後取出。
3. 將豬絞肉分為 6 等份，用手搓成長圓形，並包成高麗菜卷，最後以牙籤固定開口。
4. 小鍋內煮滾調味料，放入高麗菜卷、紅蘿蔔及鴻喜菇，加蓋燜煮 20 分鐘。

Procedures
1. Thickly slice carrot, mold into flower shape. Remove root from shimeji mushroom, wipe.（Flower shape carrot for better presentation.）
2. Cook cabbage in boiling water till tender.
3. Divide minced pork into 6 pcs. Roll meat paste into shape of long oval. Wrap by cabbage.
4. Boil seasoning in a small pot. Add cabbage rolls, shimeji mushroom and carrot, cover and simmer for 20 minutes.

料理小提示 Cooking Tips

把少許湯汁一同盛盤，可使高麗菜肉卷不容易變乾。
Serve together with soup to keep the moisture of cabbage rolls.

高麗菜肉卷是日本人很愛的家常菜，簡單的日式高湯點綴高麗菜的清新，帶出獨特的和風味道。

中華料理風的日式家常菜，跟我們的咕咾肉相類似，但肉丸內加有蔬菜，令味道更豐富。

咕咾肉丸子

便當

Meat ball in Sweet and Sour Sauce

材料
已準備的豬絞肉 150 克
洋蔥 1/2 個
玉米粉 2 湯匙

調味料
番茄醬 2 湯匙
米醋 1 湯匙
糖 2 茶匙

Ingredients
Prepared Minced Pork 150g
Onion 1/2 pcs
Corn Starch 2 tbsp

Seasoning
Ketchup 2 tbsp
Vinegar 1 tbsp
Sugar 2 tsp

作法
1. 洋蔥切小塊；豬絞肉用手搓成約 2.5 公分（1 吋）大小的肉丸。
2. 小鍋內放油燒熱，肉丸先沾上玉米粉，再炸至金黃及浮起。
3. 小碗內拌勻調味料。
4. 平底鍋放油燒熱，加入洋蔥炒至軟，倒入作法 3 拌好的調味料及炸肉丸炒勻及收汁。

Procedures
1. Dice Onion. Roll minced pork into shape of balls.
2. Coat meat balls with corn starch. Deep fry in boiling oil till golden brown. Set aside.
3. Mix seasoning in a small bowl.
4. Heat oil in frying pan, stir fry onion till tender. Add meat balls and seasoning, cook till sauce thicken.

 料理小提示 Cooking Tips

可以隨喜好加入鳳梨及青椒一起炒，增加口感之餘，也讓顏色更吸引人。
Stir with pineapple and bell pepper according to your preference.

晚餐　便當

豬肉大根煮

根菜炒豬肉

兩餐前準備 Preparation

材料
薄切嫩肩豬肉片 500 克

Ingredients
Thinly Sliced Pork Tenderloin 500g

醃料
日式醬油 2 湯匙
味醂 1 湯匙
料理酒 1 湯匙
薑汁 1 湯匙

Marinade
Japanese Soy Sauce 2 tbsp
Mirin 1 tbsp
Sake 1 tbsp
Ginger Sauce 1 tbsp

 作法 Procedures

大碗內將薄切嫩肩豬肉片及醃料拌勻。
Mix pork tenderloin slice with marinade in bowl.

 豬肉大根煮
Braised Pork Tenderloin with Radish

根菜炒豬肉 ——
Stir-fry Pork Tenderloin with Root Vegetables

晚餐

豬肉大根煮

Braised Pork Tenderloin with Radish

材料
已準備的薄切嫩肩豬肉片 250 克
大根（白蘿蔔）1/3 個
小辣椒 1 支

Ingredients
Prepared Thinly Sliced Pork Tenderloin 250g
Radish 1/3 pcs
Chili 1 pc

調味料
日式醬油 2 湯匙
味醂 3 湯匙
日式高湯 500 毫升

Seasoning
Japanese Soy Sauce 2 tbsp
Mirin 3 tbsp
Japanese Broth 500ml

作法
1. 大根去皮，切厚片，再切成 2 公分厚的長條狀。
2. 小鍋內煮滾調味料，放入大根以小火煮約 20 分鐘。（先煮大根，可令大根更熟更入味。）
3. 放入嫩肩豬肉片及小辣椒，煮至滾及豬肉片熟透。

Procedures
1.Peel radish, thickly slice, cut into 2cm width stick.
2.Bring seasonging to boil. Add radish and simmer for 20 minutes.
　（ It will make radish more tasty.）
3.Stir in pork tenderloin slice and chili, cook till pork done.

 料理小提示 Cooking Tips

不嗜辣的，可以不加小辣椒；嗜辣的，吃時可再搭配七味粉。
Omitted the chili if you do not like spicy food. Serve with
Japanese chili powder if you are a fan of spicy food.

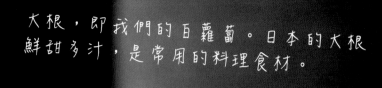

大根，即我們的白蘿蔔。日本的大根
鮮甜多汁，是常用的料理食材。

很有口感的一道料理，有肉又有蔬菜，
營養價值也很高呢！

 根菜炒豬肉

Stir-fry Pork Tenderloin with Root Vegetables

材料
已準備的薄切嫩肩豬肉片 250 克
茄子 1 條
蓮藕 1 節
白芝麻 1 茶匙

調味料
鹽 1/2 茶匙
料理酒 1 湯匙
糖 1 茶匙

Ingredients
Prepared Thinly Sliced Pork Tenderloin 250g
Eggplant 1 pc
Lotus Root 1 section
Sesame 1 tsp

Seasoning
Salt 1/2 tsp
Sake 1 tbsp
Sugar 1 tsp

作法
1. 茄子切粗條；蓮藕切薄片後再切半。
   （如果怕蓮藕變色，可將其泡在加了醋的水內。）
2. 平底鍋加油燒熱，放茄子及蓮藕炒至軟，加鹽調味。
3. 放入嫩肩豬肉片、糖及料理酒，快炒至豬肉片熟透及收汁。
4. 盛起並撒上白芝麻。

Procedures
1. Roughly shred eggplant. Thinly slice lotus root, cut slice into half.
   （Place lotus root in water with vinegar stop lotus root from going brown.）
2. Heat oil in frying pan, stir fry eggplant and lotus root till tender, season with salt.
3. Add pork tenderloin slice, sake and sugar. Stir over high heat till sauce thicken.
4. Dish up. Drizzle with sesame.

 料理小提示 Cooking Tips

配上數塊蛋卷，一點點醃菜，就變成了很和式的便當。
Together with rolled egg and pickles, make a perfect Japanese style lunchbox.

還記得，初出來社會工作時，我已經愛帶便當。
部門的同事外出吃午飯，
我自己一個人躲在茶水間加熱便當。
那是外婆給我做的午餐盒，沒有新做的飯菜，
只有前一天晚上的剩菜剩飯。
沒有精緻的裝飾，也沒有搭配的色彩；
卻是世界上最美味的午飯。

便當盒內的每一口肉，每一口菜，都由外婆親手處理烹調；
外婆用心煮出來的飯菜，帶著任何人都模仿不來的味道。
能夠在家以外，吃著家的味道，是最幸福的事。

我希望，我能夠把這種幸福帶給我的家人；
在繁忙的辦公室，吃著溫暖的便當，讀著我給你的信息：
「身邊，總有我為你打氣。」

Chapter 3
牛肉料理

具有飽足感的牛肉,可變化出漢堡排、牛肉卷、壽喜燒
等多樣料理,午餐時間打開便當會有意想不到的驚喜!

晚餐　便當

日式漢堡排

焗日式咖哩飯

兩餐前準備 Preparation

材料
牛絞肉 300 克
豬絞肉 200 克

Ingredients
Ground Beef 300g
Minced Pork 200g

作法 Procedures

大碗內將牛絞肉及豬絞肉徹底混合，成為混合肉。
Mix ground beef and minced pork in a big bowl.

焗日式咖哩飯
Baked Japanese Curry Rice

日式漢堡排 *Japanese Burger Steak*

焗日式咖哩飯

Baked Japanese Curry Rice

材料
已準備的混合肉 250 克
洋蔥 1/2 個
青椒 1 個
奶油 2 湯匙
披薩用起司絲 適量
白飯 2 碗

調味料
日式咖哩塊 1 小塊
糖 1 茶匙
水 50 毫升

Ingredients
Prepared Mixed Meat 250g
Onion 1/2 pcs
Bell Pepper 1 pc
Butter 2 tbsp
Shredded Pizza Cheese little
Rice 2 bowls

Seasoning
Japanese Curry Paste 1 pc
Sugar 1 tsp
Water 50g

作法
1. 洋蔥切碎，青椒切小粒。
2. 小鍋內放日式咖哩塊，加水煮至溶化，並加入糖拌勻，關火備用。
3. 平底鍋加奶油燒熱，放洋蔥碎炒至軟，加入青椒及混合肉以大火炒熟。
   （奶油能夠提升整個咖哩的味道層次。）
4. 加入作法 2 的咖哩醬汁，炒至略微收汁。
5. 焗烤盤內盛入白飯，加上咖哩牛肉，再平均鋪上披薩起司絲，放進烤箱
   以 180℃烤至起司融化。

Procedures
1. Finely chop onion. Dice bell pepper.
2. Mix Japanese curry paste and water in a small pot, bring to boil. Add sugar, mix well. Set aside.
3. Heat frying pan with butter, stir fry onion till tender. Add bell pepper and mix meat, stir fry over high heat till done.（Butter provides extra favor to curry.）
4. Add curry sauce, cook until sauce thicken.
5. Place rice in baking tray, add curry beef on top. Drizzle with shredded pizza cheese. Bake at 180℃ till cheese melted.

 料理小提示 Cooking Tips
可以搭配蔬菜沙拉一起享用。
A bowl of veggie salad would be best match to the dish.

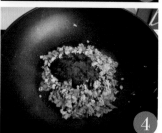

日式咖哩飯很美味，但很費時。工作一整天已累得要命了，又想吃可口的晚餐？做個簡易版焗咖哩飯也不錯呀！

經典日式西菜，也是最受歡迎的便當菜式，
聽說誰都無法抗拒漢堡排的魔力。

日式漢堡排

Japanese Burger Steak

材料	Ingredients
已準備的混合肉 250 克	Prepared Mixed Meat 250g
洋蔥 1/2 個	Onion 1/2 pcs
雞蛋 1 顆	Egg 1 pc
麵包粉 50 克	Bread Crumbs 50g
披薩用起司絲 2 湯匙	Shedded Pizza Cheese 2 tbsp

調味料	Seasoning
日式醬油 2 湯匙	Japanese Soy Sauce 2 tbsp
味醂 1 湯匙	Mirin 1 tbsp
料理酒 1 湯匙	Sake 1 tbsp
玉米粉 1 湯匙	Corn Starch 1 tbsp
胡椒粉 1 茶匙	Pepper 1 tsp

作法

1. 洋蔥切碎。平底鍋放少許油，放洋蔥炒至軟，盛入大碗內備用。
   （洋蔥必須先炒熟，才能帶出獨有的香味。）
2. 待洋蔥放涼後，加入混合肉、雞蛋、麵包粉、起司絲及所有調味料，用手拌勻。
3. 將肉分為 4 等份，用手將其中一份搓成圓餅，重複搓出其餘 3 個肉餅。
4. 完成的肉餅放至已燒熱的平底鍋上，用大火煎上下兩面，再加水 2 湯匙，蓋上蓋，轉小火煮約 5 分鐘。

Procedures

1. Finely chop onion. Heat frying pan with oil, stir fry onion till tender. Set aside.
   （Onion must be cooked before mixed with other ingredients. It allows a special favor to burger steak.）
2. After onion cooled down. Mix onion with mixed meat, egg, bread crumbs, cheese and seasoning.
3. Divide the meat paste into 4 equal portion, roll meat paste into shape of oval cake.
4. Heat frying pan with oil, pan fry burger steak till both sides brown. Add 2 tbsp of water, cover and simmer for 5 minutes.

 料理小提示 Cooking Tips

吃時可以搭配日式烤肉醬或番茄醬，酸酸甜甜。日式烤肉醬在各大日系超市皆有販售。
Serve with either Japanese BBQ sauce or ketchup. Japanese BBQ sauce available in Japanese supermarkets.

晚餐　便當

汁燒蔬菜
牛肉卷

牛丼

兩餐前準備 Preparation

材料
雪花牛肉片 300 克

Ingredients
Thinly Slice Beef Tenderloin 300g

作法 Procedures

先挑選出 8 片較完整的牛肉片，用來做汁燒蔬菜牛肉卷。
Choose 8 pcs of beef tenderloin slice which are suitable for making beef roll.

└ 🍲 牛丼 *Teriyaki Beef Bowl*

└ 🍲 汁燒蔬菜牛肉卷 *Beef Roll with Assorted Vegetables*

 牛丼

Teriyaki Beef Bowl

材料
雪花牛肉片 200 克
洋蔥 1/2 個
蔥 1 株
白飯 2 碗

Ingredients
Thinly Slice Beef Tenderloin 200g
Onion 1/2 pcs
Green Onion 1 stalk
Rice 2 bowls

調味料
日式醬油 2 湯匙
味醂 1 湯匙
料理酒 1 湯匙
糖 1 茶匙
鰹魚粉 1 茶匙

Seasoning
Japanese Soy Sauce 2 tbsp
Mirin 1 tbsp
Sake 1 tbsp
Sugar 1 tsp
Daishi Powder 1 tsp

作法
1. 洋蔥切絲；蔥切小粒；調味料放小碗內攪勻。
2. 平底鍋放油燒熱，以中小火將洋蔥炒至軟。（洋蔥在大火烹調下容易變焦糊。）
3. 加入牛肉片，並加入拌好的調味料，以大火將牛肉片炒熟及收汁。
4. 牛肉放在飯上，吃時搭配少許蔥花。

Procedures
1.Thinly slice onion. Finely chop green onion. Mix seasoning in a small bowl.
2.Heat frying pan with oil, stir fry onion over low to medium heat.
　（Onion will easily get burn if cook over high heat.）
3.Add beef tenderloin slice and seasoning. Stir fry till sauce thicken.
4.Place beef on top of rice, serve with green onion.

 料理小提示 Cooking Tips

嗜辣的，可以加少許七味粉，滋味更誘人。
Drizzle with Japanese chili powder for a spicy favor.

在日本，牛丼是丈夫的專屬料理，簡單
美味，給妻子一點點偷閒的時間。

新鮮爽口的蔬菜，給肥牛片帶來清新的
味道，是喜愛肉食料理者的健康選擇。

汁燒蔬菜牛肉卷

Beef Roll with Assorted Vegetables

材料
雪花牛肉片 8 片
紅蘿蔔 1 條
玉米筍 8 根
金針菇 1 把

Ingredients
Thinly Slice Beef Tenderloin 8 pcs
Carrot 1 pc
Baby Corn 8 pcs
Enokitake Mushroom 1 pack

調味料
日式醬油 2 湯匙
味醂 1 湯匙
料理酒 1 湯匙
糖 1 茶匙

Seasoning
Japanese Soy Sauce 2 tbsp
Mirin 1 tbsp
Sake 1 tbsp
Sugar 1 tsp

醬汁
日式醬油 1 湯匙
味醂 1 湯匙
水 100 毫升
鰹魚粉 1 茶匙

Sauce
Japanese Soy Sauce 1 tbsp
Mirin 1 tbsp
Water 100ml
Daishi Powder 1 tsp

作法
1. 紅蘿蔔切細條；玉米筍切半；金針菇去蒂。
2. 小鍋內煮滾醬汁，加紅蘿蔔煮至軟、玉米筍及金針菇略熟。
3. 調味料放小碗內拌勻。
4. 紅蘿蔔、玉米筍及金針菇平均分成 8 份，取牛肉片將一份捲起，重複完成 8 份牛肉卷。
5. 平底鍋加油燒熱，放入牛肉卷煎至八成熟，拌入調味料，煮至收汁。
（必須先從開口部位開始煎，才能令牛肉卷不鬆散。）

Procedures

1. Shred carrot. Cut baby corn in half. Remove root from enokitake mushroom.
2. Add sauce in a small pot, boil carrot, baby corn and enokitake till tender.
3. Mix seasoning in a small bowl.
4. Divide the vegetables into 8 equal portions. Roll up one portion of vegetable by beef tenderloin slice.
5. Heat frying pan with oil, pan fry beef roll till 80% cooked. Pour in seasoning, cook till sauce thicken. （Cook the open end of the beef roll first.）

 料理小提示 Cooking Tips

牛肉卷內已有大量的蔬菜，味道也較濃烈，可以在便當內搭配較清新爽口的配菜，添加不同的口感。
The beef roll comes with lots of vegetable in rich taste. It would be perfectly match with light tasted food in lunchbox.

晚餐　　便當

吉列
牛肉薯餅

和風
春雨沙拉

兩餐前準備 Preparation

材料
牛絞肉 300 克
蒜蓉 1 茶匙

調味料
日式醬油 1 湯匙
料理酒 1 茶匙
麻油 1 茶匙

Ingredients
Ground Beef 300g
Minced Garlic 1 tsp

Seasoning
Japanese Soy Sauce 1 tbsp
Sake 1 tsp
Sesame Oil 1 tsp

 作法 Procedures

1. 平底鍋加油燒熱，爆香蒜蓉。
2. 加入牛絞肉略炒至轉色，加調味料大火炒至收汁。
1. Heat frying pan with oil, saute minced garlic.
2. Add ground beef, stir fry. Add seasoning, stir fry till sauce thicken.

和風春雨沙拉
*Japanese Style Beef and
Vermicelli Salad*

吉列牛肉薯餅 ——
Beef Croquet

和風春雨沙拉

Japanese Style Beef and Vermicelli Salad

材料
已準備的牛絞肉 200 克
春雨（粉絲）30 克
紅蘿蔔 1 條
小黃瓜 1 條
白芝麻 1 茶匙

Ingredients
Prepared Ground Beef 200g
Japanese Vermicelli 30g
Carrot 1 pc
Cucumber 1 pc
Sesame 1 tsp

調味料
日式醬油 1 湯匙
糖 1 茶匙
七味粉 1 茶匙

Seasoning
Japanese Soy Sauce 1 tbsp
Sugar 1 tsp
Japanese Chili Powder 1 tsp

作法
1. 紅蘿蔔及小黃瓜去皮，切絲。春雨以熱水浸泡 5 分鐘，瀝乾備用。
2. 平底鍋放少許油燒熱，加紅蘿蔔及小黃瓜絲炒至軟，加少許鹽調味，盛起。
3. 在同一個平底鍋內加入牛絞肉及泡過的春雨，翻炒至香氣出來，加入紅蘿蔔及小黃瓜絲，拌入調味料，大火炒至收汁。
4. 盛起放涼，吃時撒上白芝麻。

Procedures
1.Peel carrot and cucumber, shred. Soak vermicelli for 5 minutes, drain well.
2.Heat frying pan with oil, stir fry carrot and cucumber till tender, season with salt. Set aside.
3.Add ground beef and vermicelli in the same frying pan, stir fry. Add carrot, cucumber and seasoning, stir fry over high heat till sauce thicken.
4.Dish up. Drizzle with sesame when serve.

 料理小提示 Cooking Tips

這道沙拉可以當主菜吃，也可以當配菜吃，看看胃口怎樣再打算就好了！
It can serve as either appetizer or main dish, according to your preference.

春雨就是我們的粉絲，一樣由綠豆製成。
但日本的春雨口感更有嚼勁，也不容易
過熟，是做沙拉的好選擇！

胖嘟嘟的薯餅很惹人喜愛，多做幾個當小吃也可以呀！

吉列牛肉薯餅

Beef Croquet

材料

已準備的牛絞肉 100 克
馬鈴薯 2 個
鮮奶 2 湯匙
玉米粉 2 湯匙
雞蛋 1 顆
麵包粉 1 碗

Ingredients

Prepare Ground Beef 100g
Potato 2 pcs
Milk 2 tbsp
Corn Starch 2 tbsp
Egg 1 pc
Bread Crumbs 1 bowl

調味料

鹽 1 茶匙
胡椒粉 1 茶匙

Seasoning

Salt 1 tsp
Pepper 1 tsp

作法

1. 馬鈴薯去皮切大塊。小鍋內燒滾水，放馬鈴薯煮 20 分鐘至熟。
2. 馬鈴薯放入大碗內，用叉子壓成泥，加入牛絞肉及鮮奶拌勻，並以鹽和胡椒粉調味。
3. 將薯泥分成 4 等份，分別用手將薯泥搓成雞蛋形薯餅。將薯餅依序沾上玉米粉、雞蛋及麵包粉。（也可以按喜好做成不同形狀或大小。）
4. 大鍋裡燒滾油，放入薯餅炸至金黃。

Procedures

1. Peel potato, cut into big pieces. Add water in a small pot, bring potato to boil for 20 minutes till tender.
2. Mash boiled potato in a big bowl, mix in ground beef and milk, season with salt and pepper.
3. Divide mashed potato into 4 equal portions, roll mashed potato into shape of oval cake. Coat potato cake with corn starch, beaten egg and bread crumbs. （You may change shape or size of croquet according to your preference.）
4. Deep fry potato cake in boiling oil till golden brown.

> 料理小提示 Cooking Tips
>
> 這道薯餅可以加進不同種類的剩肉一起製作。
> You can change to any kind of meat according to your preference.

晚餐　便當

和風牛肉豆腐湯　豆腐壽喜燒

兩餐前準備 Preparation

材料
牛肋條 400 克
日式高湯 600 毫升

Ingredients
Beef Rib 400g
Japanese Broth 600ml

作法 Procedures

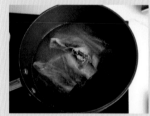

1. 牛肋條放小鍋中汆燙，盛起。
2. 將牛肋條切成 2 ～ 3 公分的長度。
3. 小鍋內再放入日式高湯煮滾，加入已汆燙的牛肋條以小火燜煮約 1 小時。

1. Rinse and scalding beef rib in boiling water for a short while. Rinse again, set aside.
2. Cut beef rib into 2-3cm long pieces.
3. Boil Japanese broth in a pot, braise over low heat for 1 hour.

 料理小提示 Cooking Tips

這兩道菜用到的高湯比較多，建議可直接製作日式高湯使用，作法可參考 p.9。
Since this menu consists of large amount of broth, suggested to make broth rather than instant dashi powder. Please refer to page 9.

和風牛肉豆腐湯
Japanese Braised Beef and Tofu Soup

豆腐壽喜燒 ——
Sukiyaki with Tofu

和風牛肉豆腐湯

Japanese Braised Beef and Tofu Soup

材料

已準備牛肋條 200 克
燜煮牛肋條的高湯 400 毫升
豆腐 1 塊
鮮冬菇 6 顆
巴西里 少許

調味料

日式醬油 2 湯匙
味醂 2 湯匙
料理酒 2 湯匙

Ingredients

Prepared Braised Beef Rib 200g
Japanese Broth from Preparation 400ml
Hard Tofu 1 pc
Fresh Shitake Mushroom 6 pcs
Parsley little

Seasoning

Japanese Soy Sauce 2 tbsp
Mirin 2 tbsp
Sake 2 tbsp

作法

1. 豆腐切厚片，鮮冬菇用小刀在菇面劃出十字形。
   （在鮮冬菇上劃十字，只是為了美觀，怕麻煩者可省去此步驟。）
2. 瀝去牛肋條高湯多餘的油脂。
3. 小鍋內倒入高湯及牛肋條煮滾，加入豆腐、鮮冬菇及調味料，再以小火煮約 15 分鐘。
4. 吃時可搭配巴西里。

Procedures

1. Dice tofu into big pieces. Cut a cross on shitake mushroom.
   （The cross on shitake mushroom for better presentation.）
2. Drain excess oil and fat from Japanese broth.
3. Add broth and braised beef rib in pot, bring to boil. Add tofu, mushroom and seasoning, simmer for 15 minutes.
4. Serve with Parsley.

 料理小提示 Cooking Tips

在燜煮牛肋條期間，可以預備其他材料，節省製作時間。
Prepare other ingredients when the beef rib in braising.

牛肋條的料理往往較費時。
回家後，先開始準備高湯及
牛肋條，待牛肋條在燜煮時，
就很多時間閒著了。

壽喜燒即大家熟識的 Sukiyaki，是宴客的菜式呢。把它放進便當盒，滿滿的幸福感覺！

便當

豆腐壽喜燒
Sukiyaki with Tofu

材料
已準備的牛肋條 200 克
油豆腐 2 塊
洋蔥 1/2 個

調味料
日式醬油 1 湯匙
味醂 1 湯匙
料理酒 1 湯匙
糖 2 茶匙

Ingredients
Prepared Braised Beef Rib 200g
Deep Fried Tofu 2 pcs
Onion 1/2 pcs

Seasoning
Japanese Soy Sauce 1 tbsp
Mirin 1 tbsp
Sake 1 tbsp
Sugar 2 tsp

作法
1. 豆腐切厚片；洋蔥切絲。
2. 平底鍋放油燒熱，將豆腐煎至兩面呈輕微金黃色，盛起。
   （豆腐煎過後，不易散開。）
3. 在同樣的鍋子內放洋蔥炒至軟。
4. 加入牛肋條及豆腐，拌入調味料，以大火炒至收汁。

Procedures
1. Thickly slice deep fried tofu. Shred onion.
2. Heat frying pan with oil, pan fry tofu till both sides golden brown. Set aside.
   （Tofu is not easy to break after pan fried.）
3. Stir fry onion in the same frying pan till tender.
4. Add beef rib, tofu and seasoning. Stir fry over high heat till sauce thicken.

料理小提示 Cooking Tips

可以配上蛋卷及一點點的鮮豔蔬菜，增加華麗的感覺。
It makes a perfect lunchbox with homemade egg roll and colorful vegetables.

晚餐　便當

青椒牛肉絲

味噌牛肉
菇菇寬麵

兩餐前準備 Preparation

材料	醃料
牛柳 2 塊	日式醬油 2 湯匙
洋葱 1 個	味醂 2 湯匙
	料理酒 2 湯匙
	糖 2 茶匙
	水 3 湯匙

Ingredients	Marinade
Beef Tenderloin 2 pcs	Japanese Soy Sauce 2 tbsp
Onion 1 pc	Mirin 2 tbsp
	Sake 2 tbsp
	Sugar 2 tsp
	Water 3 tbsp

 作法 Procedures

1. 牛肉切條，加醃料拌勻，醃至少 20 分鐘。
2. 洋葱切絲。
3. 平底鍋加少許油燒熱，放洋葱炒至軟，加入牛肉快炒至八分熟，盛起。

1.Slice beef tenderloin. Mix with marinade, set aside for at least 20 minutes.
2.Finely shred onion.
3.Heat frying pan with oil, stir fry onion till tender. Add beef tenderloin, stir fry till 80% cooked.

味噌牛肉菇菇寬麵
*Linguine with Tenderloin
and Assorted Mushroom*

 青椒牛肉絲 ——
*Stir-fry Slice Tenderloin
and Green Pepper*

晚餐

味噌牛肉菇菇寬麵
Linguine with Tenderloin and Assorted Mushroom

材料
已準備的牛柳及洋蔥絲 1/2 份
鴻喜菇 1 盒
小辣椒 1 支
蒜蓉 1 茶匙
義大利寬麵 200 克
巴西里 少許

調味料
味噌 2 茶匙
糖 2 茶匙
料理酒 1 湯匙
熱水 3 湯匙

Ingredients
Prepared Tenderloin and Onion 1/2 portion
Shimeji Mushroom 1 pack
Chili 1 pc
Minced Garlic 1 tsp
Linguine 200g
Parsley little

Seasoning
Miso 2 tsp
Sugar 2 tsp
Sake 1 tbsp
Hot water 3 tbsp

作法
1. 義大利寬麵放入滾水內,加鹽,按照包裝袋上的時間指示烹煮,瀝乾盛盤備用。
   (寬麵可以用少許橄欖油拌和,及蓋上濕毛巾,防止變乾。)
2. 鴻喜菇去蒂頭尾部,小辣椒切絲,調味料放小碗內拌勻。
3. 平底鍋放油燒熱,放入鴻喜菇及小辣椒,炒至鴻喜菇變軟。
4. 加入牛柳、洋蔥及調味料,大火炒至收汁。
5. 將牛柳放在煮好的寬麵上,以巴西里裝飾。

Procedures
1. Cook linguine according to instruction on package. Drain.
   (Mix cooked linguine with olive oil or cover by wet towel, prevent to dry out.)
2. Remove root from shimeji mushroom. Slice chili. Mix seasoning in small bowl.
3. Heat frying pan with oil. Stir shimeji mushroom and chili till shimeji mushroom tender.
4. Add beef tenderloin , onion and seasoning, stir fry over high heat till sauce thicken.
5. Place beef tenderloin on linguine, serve with parsley.

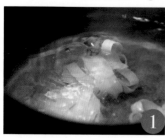  

 料理小提示 Cooking Tips

味噌本身帶有鹹味,故不建議另外加鹽,若覺得味噌味太濃,可以增加糖的分量調和一下。
Miso carries very salty taste, it is not necessary to add extra salt. Sugar helps to neutralize miso if too salty.

日式味噌跟義大利寬麵，像是風馬牛不相及的兩回事，但卻出奇地合襯，讓人再三回味。

不用5分鐘就可以完成的
便當料理，可以多點時間
賴床了！

青椒牛肉絲

Stir-fry Slice Tenderloin and Green Pepper

材料
已準備的牛柳及洋葱絲 1/2 份
青椒 1 個
紅蘿蔔 1/2 條

Ingredients
Prepared Tenderloin and Onion 1/2 portion
Bell Pepper 1 pc
Carrot 1/2 pcs

調味料
日式醬油 1 湯匙
味醂 1 湯匙
料理酒 1 湯匙
糖 2 茶匙
薑汁 1 湯匙

Seasoning
Japanese Soy Sauce 1 tbsp
Mirin 1 tbsp
Sake 1 tbsp
Sugar 2 tsp
Ginger Sauce 1 tbsp

作法
1. 青椒、紅蘿蔔切絲。
2. 調味料放小碗內拌勻。
3. 平底鍋放油燒熱，加入青椒及紅蘿蔔絲炒至稍軟。
4. 加入牛柳及洋葱絲，拌入調味料，快炒至收汁。

Procedures
1. Shred bell pepper and carrot.
2. Mix seasoning in a small bowl.
3. Heat frying pan with oil, stir fry bell pepper and carrot till tender.
4. Add beef tenderloin, onion and seasoning. Stir fry over high heat till sauce thicken.

料理小提示 Cooking Tips

這道料理的湯汁較多，把肉放在米飯上面，香濃的湯汁跟米飯混合，對著這道美味的料理，一個午餐盒真的足夠嗎？
Place the cooked beef tenderloin over rice, let the sance mix together with rice. That would be the best choice for bustle lunch time.

從小生長在一個惜食的家庭。
家裡的剩菜剩飯，總會留到下一餐加熱了再吃。
長輩也從不容許我的碗盤內留下一小口的食物。
他們總是說：
要知道世界上很多孩子連白米飯也沒有見過，
我能夠有吃的，已經是很大的福氣。

童年的教誨，成為了今天的習慣，
珍惜食物這個道理，根深柢固地種在我的心裡。
對每一口食物有感恩之心，感受天地賜予的珍貴。
從惜食到知足，讓我明白生活的快樂及美好。

惜食，原來是一種幸福。

Chapter 4
海鮮料理

運用鰻魚、鯖魚、鮭魚……等富含營養的海鮮食材，做
出鮮味十足的料理，讓人吃了之後充滿朝氣與活力。

晚餐 鰻魚柳川鍋

便當 鰻魚炒飯

兩餐前準備 Preparation

材料	醃料	Ingredients	Marinade
蒲燒鰻魚 2 條	料理酒 2 湯匙	Roasted Eel 2 pcs	Sake 2 tbsp

🍴 作法 Procedures

1. 預備錫箔紙 2 張。
2. 將其中一條鰻魚切大塊,另一條切小塊。
3. 切好後分別放在不同的錫箔紙上,灑上料理酒,包裹。
4. 將 2 包鰻魚一同放入大鍋中蒸 10 分鐘。

1.Prepare 2 pcs of foil.
2.Cut one of the roasted eel in bite size, while finely dice another roasted eel.
3.Place on 2 different piece of foil, drizzle with sake. Wrap.
4.Steam wrapped roasted eel for 10 minutes.

鰻魚柳川鍋
Roasted Eel with Egg in Pot

鰻魚炒飯 *Fried Rice with Roasted Eel*

鰻魚柳川鍋
Roasted Eel with Egg in Pot

材料
已準備的大塊鰻魚
洋葱 1/2 個
雞蛋 1 顆
巴西里 少許

Ingredients
Prepared Roasted Eel in Bite Size
Onion 1/2 pcs
Egg 1 pc
Parsley little

調味料
日式醬油 1 湯匙
味醂 1 湯匙
料理酒 1 湯匙
糖 2 茶匙
日式高湯 100 毫升

Seasoning
Japanese Soy Sauce 1 tbsp
Mirin 1 tbsp
Sake 1 tbsp
Sugar 2 tsp
Japanese Broth 100ml

作法
1. 洋葱切細絲。小碗內打發雞蛋。
2. 平底鍋放油燒熱，加入洋葱炒軟。
3. 將鰻魚排放在平底鍋內，倒入調味料，蓋上，煮至滾。
4. 倒入已打發的雞蛋，輕輕推開，煮至蛋熟。
　（不要經常攪拌雞蛋，才能保持整個鍋的美觀。）
5. 吃時搭配巴西里。

Procedures
1.Shred onion. Beat egg in a small bowl.
2.Heat oil in frying pan, stir fry onion till tender.
3.Add roasted eel and seasoning. Cover and simmer till boiled.
4.Add beaten egg, gently spread thru whole pan. Cook till egg done.
　（Do not over stir egg.）
5.Serve with parsley.

 料理小提示 Cooking Tips

日本人愛將牛蒡加進這道料理之內，但牛蒡不常使
用，故改用洋葱，減少浪費。
The original recipe use burdock. Since burdock is not
popular, so we change to onion.

甜甜的湯汁，加上濃濃的鰻魚，就算晚餐就這麼一道菜，也可以吃上兩大碗飯。

一個電鍋就可以做到的鰻魚炒飯，把早上的繁忙節奏放鬆下來，騰出時間來杯醒神咖啡！

鰻魚炒飯

Fried Rice with Roasted Eel

材料
已準備的小塊鰻魚
蓮藕 1 節
珍珠米 2 杯

Ingredients
Prepared Diced Roasted Eel
Lotus Root 1 section
Uncooked Rice 2 cup

醬汁
日式醬油 3 湯匙
味醂 2 湯匙
糖 2 茶匙
日式高湯 450 毫升

Sauce
Japanese Soy Sauce 3 tbsp
Mirin 2 tbsp
Sugar 2 tsp
Japanes Broth 450ml

調味料
蒲燒汁 2 湯匙

Seasoning
Japanese Eel Sauce 2 tbsp

作法
1. 蓮藕切小粒。小鍋內煮滾日式高湯，放蓮藕粒煮 15 分鐘。
2. 米洗淨，連同高湯及蓮藕粒一起放進電鍋內，啟動正常煮飯模式。
3. 蓮藕飯煮好後，放入鰻魚塊及蒲燒汁拌勻。
4. 蓋上鍋蓋，啟動翻熱模式。(如電鍋沒有翻熱模式，可再煮約5～8分鐘。)
5. 可加上海苔碎及柴魚片一同享用。

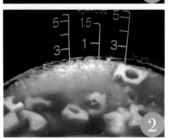

Procedures

1. Dice lotus root. Add sauce and lotus root in a small pot, bring to boil for 15 minutes.
2. Rinse rice. Cook rice, lotus root and sauce in rice cooker.
3. When rice done, add roasted eel and seasoning. Stir well.
4. Cover, turn on Reheat mode.
   (If no reheat mode available, cook rice for another 5-8 minutes.)
5. Serve with shredded seaweed or bonito flakes.

料理小提示 Cooking Tips

記得在便當盒內放點日式醃蘿蔔，跟炒飯最搭配。
Do not forget a piece of radish pickle in lunchbox.

晚餐　鯖魚味噌煮

便當　鯖魚板壽司

兩餐前準備 Preparation

材料	醃料	Ingredients	Marinade
半邊的鯖魚 4 份	鹽 2 茶匙	Half Cut Mackerel 4 pcs	Salt 2 tsp

作法 Procedures

1. 鯖魚洗淨後把水擦乾。
2. 將其中 2 份鯖魚分別切片成 3 等份。
3. 平底鍋燒熱，鯖魚入鍋（帶皮面朝下），煎至皮脆。
4. 翻面再煎至另一面全熟，加鹽調味。

1.Rinse mackerel. Wipe dry.
2.Cut 2 pcs of mackerel into 3 sections each.
3.Heat frying pan with oil, turn the side with skin down to the pan. Pan fry till skin turned crispy.
4.Turn over and pan fry another side. Seasoning with salt.

 鯖魚味噌煮 *Mackerel Simmered in Miso*

鯖魚板壽司
Mackerel Sushi

鯖魚味噌煮

Mackerel Simmered in Miso

材料
已準備的切片鯖魚 2 份
薑 3 片
葱 1 株
味噌 1 湯匙

調味料
日式醬油 1 湯匙
味醂 1 湯匙
料理酒 1 湯匙
糖 2 茶匙
水 150 毫升
鰹魚粉 1 茶匙

Ingredients
Prepared Mackerel in Sections 2 pcs
Ginger 3 slices
Green Onion 1 stalk
Miso 1 tbsp

Seasoning
Japanese Soy Sauce 1 tbsp
Mirin 1 tbsp
Sake 1 tbsp
Sugar 2 tsp
Water 150ml
Dashi Powder 1 tsp

作法
1. 薑切絲；葱切小段。
2. 小鍋放油燒熱，炒香薑絲及葱段。
3. 下調味料煮滾。
4. 放入鯖魚片，待湯汁煮滾後。再用篩網將味噌融入湯內，稍煮後熄火。
   （味噌大滾會將香味揮發，要小心控制火候。）

Procedures
1.Shred onion. Cut green onion into sections.
2.Heat a small pot with oil, saute ginger and green onion.
3.Add seasoning, bring to boil.
4.Add mackerel. When the sauce start boiling, sieve in miso. Turn off.
   （the favor of miso will be lighten over high heat.）

 料理小提示 Cooking Tips

喜歡濃郁味噌味道的朋友，可選用赤味噌，不要一味添加大量
味噌，令料理過鹹。
Replace by red miso for a more rich miso favor.

116

鯖魚的魚油香，跟味噌的濃郁互相搭配融合，這是日本人最愛的魚料理。

日本火車便當的鯖魚板壽司，把煎過的魚
白飯上，讓濃香的魚油慢慢滲進飯內。

鯖魚板壽司
Mackerel Sushi

材料
已準備的鯖魚 2 份
煮好的珍珠米飯 2 碗
紫蘇 4 片
白芝麻 1 茶匙
巴西里 少許

Ingredients
Prepared Half Cut Mackerel 2 pcs
Rice 2 bowl
Japanese Basil Leaves 4 pcs
Sesame 1 tsp
Parsley little

調味料
米醋 3 湯匙
糖 1 湯匙
鹽 2 茶匙

Seasoning
Vinegar 3 tbsp
Suagr 1 tbsp
Salt 2 tsp

作法
1. 白飯與調味料拌勻，放涼。
2. 壽司竹卷上先鋪一層保鮮膜，放上壽司飯，捲成長飯糰狀，壓實。
3. 攤開竹卷，在飯上鋪紫蘇及鯖魚，捲起壓實。
4. 壽司切段後，每段放上巴西里及白芝麻作裝飾。

Procedures
1. Mix rice with seasoning. Set aside.
2. Place a piece of wrap over sushi mat, add rice. Roll into shape of rectangular prism. Press tight.
3. Open sushi mat. Place basil leaves and mackerel on rice. Roll and press tight.
4. Cut sushi into pieces, decorate with parsley and sesame.

🍲 料理小提示 Cooking Tips

若講究味道，品嘗時可以再淋上些許鮮檸檬汁，點綴鯖魚的魚香味。
Few drops of fresh lemon juice will elaborate the favor of mackerel.

晚餐　便當

墨魚燒餅

汁燒

香脆墨魚

兩餐前準備 Preparation

材料
墨魚 2 隻
薑 3 片
葱 1 株

Ingredients
Roasted Eel 2 pcs
Ginger 3 pcs
Green Onion 1 stalk

作法 Procedures

1. 墨魚洗淨，墨魚身體的部分切成圈狀，墨魚腳則切成 3 ～ 4 公分長。
2. 小鍋內加水煮沸，放薑及葱，將墨魚汆燙。
1. Rinse squid. Cut squid body into rings while legs into 3-4cm long pieces.
2. Boil water, add ginger and green onion. Scald squid in boiling water for a short while. Rinse again.

汁燒香脆墨魚
Teriyaki Deep Fried Squid

墨魚燒餅
Japanese Pancake with Squid

汁燒香脆墨魚
Teriyaki Deep Fried Squid

材料
已準備的墨魚 1 隻
雞蛋 1 顆
玉米粉 2 湯匙
白芝麻 1 茶匙

調味料
日式醬油 1 湯匙
味醂 1 湯匙
料理酒 1 湯匙
糖 2 茶匙

Ingredients
Prepare Squid 1 pc
Egg 1 pc
Corn Starch 2 tbsp
Sesame 1 tsp

Seasoning
Japanese Soy Sauce 1 tbsp
Mirin 1 tbsp
Sake 1 tbsp
Sugar 2 tsp

作法
1. 小碗內打發雞蛋,加玉米粉拌勻。
2. 鍋內放油燒滾,墨魚沾上蛋液後入油鍋炸至金黃,盛起。
3. 小碗內拌勻調味料。平底鍋燒熱,加入拌好的調味料煮至滾,放入墨魚大炒至收汁。
4. 吃時撒上芝麻。

Procedures
1.Beat egg in a small bowl, mix with corn starch.
2.Coat squid with beaten egg. Deep fry in boiling oil till golden. Set aside.
3.Mix seasoning in a small bowl. Heat frying pan, add seasoning and squid. Stir fry till sauce thicken.
4.Serve with sesame.

料理小提示 Cooking Tips

香氣誘人的一道料理,多做幾個小吃,簡單配上一兩杯酒水,兩口子輕鬆度過一整個晚上。
Have a relaxing night with this dish and a glass of wine.

西餐的佐酒菜有炸墨魚，家
裡的下酒菜有汁燒香脆墨
魚。週末做一道，再來一個
漫長的閒話家常也不錯吧！

日本燒餅變奏版，沒有重量醬料，卻加了輕盈美味的墨魚，以鹽調味，健康的愛心飯盒。

墨魚燒餅

Japanese Pancake with Squid

材料
已準備的墨魚 1 隻
高麗菜 1/4 個
葱 1 株

Ingredients
Prepared Squid 1 pc
Cabbage 1/4 pcs
Green Onion 1 stalk

煎餅麵糊
雞蛋 1 顆
麵粉 100 克
水 100 毫升
鹽 1 茶匙
糖 1 茶匙
鰹魚粉 1 茶匙

Okonomiyaki Batter
Egg 1 pcs
Flour 100g
Water 100ml
Salt 1 tsp
Sugar 1 tsp
Dashi Powder 1 tsp

作法
1. 高麗菜切絲。葱剁碎。
2. 大碗內加煎餅麵糊、墨魚、高麗菜絲及葱碎拌勻。
3. 平底鍋燒熱，倒入作法 2 拌好的麵糊呈圓餅狀。
4. 小火煎至金黃，翻面再煎另一面至金黃。
5. 放涼後切片，再放入便當盒內。
   （煎餅要完全放涼才入盒，可以保持香脆。）

Procedures
1.Shred cabbage. Finely chop green onion.
2.Mix batter, squid, cabbage and green onion in a bowl.
3.Heat frying pan with oil, pour batter and make a round cake.
4.Cook over low-medium heat till golden brown, turn over to another side, cook till golden brown.
5.Cool completely before cut.
（For a crispy texture, cool completely before plate in box.）

 料理小提示 Cooking Tips

可準備一些沙拉醬或海苔碎放在盒內，吃時才拌入，令煎餅更美味。
Serve with mayonnaise or shredded seaweed for a complete Japanese pancake.

晚餐 鮭魚豆腐味噌湯

便當 鹽燒鮭魚飯糰

兩餐前準備 Preparation

材料	醃料	Ingredients	Marinade
鮭魚排 2 塊	鹽 2 茶匙 胡椒粉 2 茶匙 玉米粉 2 茶匙	Salmon Steak 2 pcs	Salt 2 tsp Pepper 2 tsp Corn Starch 2 tsp

作法 Procedures

1. 鮭魚洗淨擦乾。
2. 均勻抹上鹽及胡椒粉，兩面皆拍上玉米粉。
3. 平底鍋放油燒熱，煎至鮭魚兩面金黃全熟。

1.Rinse salmon. Wipe dry.
2.Season salmon steak with salt and pepper, coat with corn starch.
3.Heat frying pan with oil, pan fry salmon steak till both sides golden brown.

鮭魚豆腐味噌湯
Miso Soup with Salmon and Tofu

鹽燒鮭魚飯糰
Fried Rice Ball
with Salmon

晚餐

鮭魚豆腐味噌湯
Miso Soup with Salmon and Tofu

材料
已準備的鮭魚排 1 塊
嫩豆腐 1 塊
葱 1 株

Ingredients
Prepared Salmon Steak 1 pc
Soft Tofu 1 pc
Green Onion 1 stalk

調味料
水 500 毫升
鰹魚粉 2 茶匙
味噌 2 湯匙

Seasoning
Water 500ml
Dashi Powder 2 tsp
Miso 2 tbsp

作法
1. 鮭魚排切成一口大小；嫩豆腐切小粒；葱切小粒。
2. 小鍋裡放水及鰹魚粉煮沸，加入豆腐，轉小火煮約 10 分鐘。
3. 放入鮭魚，煮至稍熱。（鮭魚原本已熟，再煮太久會老、肉質變硬。）
4. 味噌放入篩網內，用小湯匙將味噌融入湯內，在湯再次煮滾前關火。
5. 盛起並撒上葱花。

Procedures
1. Cut salmon steak into bite size. Dice tofu. Finely chop green onion.
2. Boil water and dashi powder in a small pot. Add tofu, simmer for 10 minutes.
3. Add salmon, boil for a short while.
   （Salmon already cooked, the boiling time in soup can be shorten.）
4. Sieve miso into soup, turn off heat before boiling.
5. Dish up. Drizzle with green onion.

 料理小提示 Cooking Tips

鮭魚先煎一下，能夠帶出其油香，令整碗味噌湯變得更美味。
TThe pan fried salmon gives miso soup a richer favor.

細火慢燉的老火湯固然濃郁美味，但簡單
的一道味噌湯，也清新而甜蜜。

白米飯吃得悶了嗎？把它搓成飯糰，加一些
新鮮感。鮭魚油融化在飯糰內，再加上燒醬
油的香，要多做幾個嗎？

便當

鹽燒鮭魚飯糰
Fried Rice Ball with Salmon

材料
已準備的鮭魚排 1 塊
白芝麻 1/2 湯匙
煮好的珍珠米飯 2 碗
壽司用海苔 1/2 塊

Ingredients
Prepared Salmon Steak 1 pc
Sesame 1/2 tbsp
Rice 2 bowls
Sushi Used Seaweed 1/2 pcs

調味料
鹽 1 茶匙
日式醬油 2 湯匙

Seasoning
Salt 1 tsp
Japanese Soy Sauce 2 tbsp

作法
1. 鮭魚放大碗內搗碎,加入白飯、白芝麻及鹽拌勻。
2. 取一張保鮮膜,將一半的飯放在保鮮膜上,包裹住及捏成三角形,將保鮮膜扭實,做成飯糰。
   重複完成另一飯糰。
3. 平底鍋燒熱,放下飯糰乾煎,過程中在兩面不停刷上醬油,及反覆煎至兩面飯焦。
   (平底鍋不要加油,乾煎的飯糰更健康香脆。)
4. 以海苔包裹飯糰再放入盒內。

Procedures
1.Mash salmon steak in a bowl, mix with rice, sesame and salt.

2.Wrap half of rice, roll in the shape of triangle rice ball. Repeat another one.

3.Heat frying pan, place rice ball. Brush on soy sauce on both side, turn occasionally.

 (Do not add oil in this step.)

4.Wrap cooked rice ball with seaweed.

 料理小提示 Cooking Tips

不愛白芝麻,可加入柴魚片,又或兩樣一起,那種獨有的口感,令人再三回味。
Relace saseme by bonito flakes or both together, according to your preference.

晚餐 便當

吉列蝦堡　海苔蝦卷

兩餐前準備 Preparation

材料	醃料	Ingredients	Marinade
蝦仁 400 克	日式醬油 1 湯匙	Prawn 400g	Japanese Soy Sauce 1 tbsp
葱 1 株	味醂 1 湯匙	Green Onion 1 stalk	Mirin 1 tbsp
	糖 1 茶匙		Sugar 1 tsp
	胡椒粉 1 茶匙		Pepper 1 tsp
	玉米粉 1 湯匙		Corn Starch 1 tbsp

作法 Procedures

1. 葱切碎。
2. 蝦仁去殼及腸泥，洗淨，剁成泥。
3. 大碗內放蝦泥、葱花及醃料拌勻。

1. Finely chop green onion.
2. Shell and devein prawn. Rinse and wipe dry. Chop prawns to paste form.
3. Mix prawn paste, onion and marinade in a bowl.

吉列蝦堡
Deep Fried Breaded Prawn Cake

海苔蝦卷
*Prawn Roll with
Seaweed*

吉列蝦堡
Deep Fried Breaded Prawn Cake

材料	Ingredients
已準備的蝦泥 200 克	Prepare Prawn Paste 200g
紅蘿蔔 1/2 條	Carrot 1/2 pcs
玉米粉 3 湯匙	Corn Starch 3 tbsp
雞蛋 1 顆	Egg 1 pc
麵包粉 1/2 碗	Bread Crumb 1/2 bowl

作法

1. 紅蘿蔔刨碎。
2. 大碗內放入蝦泥及紅蘿蔔碎，拌勻。用手搓成 5 個大小相近的圓餅。
3. 蝦餅依序沾上玉米粉、蛋液及麵包粉。
4. 小鍋中加油燒滾，放進蝦餅炸至金黃。

Procedures

1.Finely shred carrot.

2.Mix prawn paste and carrot. Divide paste into 5 equal portion. Roll prawn paste into shape of round cake.

3.Coat prawn cake with corn starch, egg and bread crumbs.

4.Deep fry prawn cake in boiling oil till golden brown.

 料理小提示 Cooking Tips

如果家裡有起司碎或起司粉，可於製作蝦餅時拌入，做出香氣撲鼻的芝心蝦餅。

For a rich favor, mix shredded cheese with prawn paste.

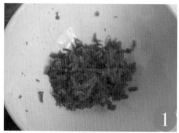

數個小蝦堡，拌點高麗菜絲沙拉、一碗熱騰騰的
白米飯和味噌湯，簡單就能夠很快樂了。

用海苔包裹再去煎的料理，日本人稱之為「磯燒」，
海苔被燒得香脆，蝦肉軟滑，真正外脆內軟。

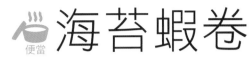 海苔蝦卷
Prawn Roll with Seaweed

材料
已準備的蝦泥 200 克
壽司用海苔 1 張

Ingredients
Prepare Prawn Paste 200g
Sushi Used Seasweed 1 pc

作法
1. 海苔剪成 10 張大小相同的小海苔。
2. 蝦泥分為 10 等份，用手搓成長餅狀，放在其中一張海苔的末端，再捲起。
3. 重複作法 2 步驟，完成其餘的蝦卷。
4. 平底鍋放油燒熱，放入蝦卷，小火煎至蝦卷熟透。
 （火太大，海苔容易變焦。）

Procedures
1.Cut seaweed into 10 equal pieces.
2.Divide prawn paste into 10 equal portions.
 Roll prawn paste into shape of oval cake. Wrap by seaweed.
3.Repeat the above, finish the entire prawn paste.
4.Heat fry pan with oil, pan fry prawn roll over low to medium heat .
（Seaweed will easily burn over high heat.）

 料理小提示 Cooking Tips

海苔顏色太深沉，在盒內放一點顏色鮮豔的蔬菜，會更漂亮。
Best match with some colorful vegetables in lunchbox.

從小跟著外婆長大。
小時候住在那種四四方方，廚房、洗手間都在窗前的舊式屋村。
家裡的廚房跟客廳，其實只有一塊玻璃之隔。
外婆在廚房裡忙，我在玻璃之外陪伴著她。
有時候專心地玩著玩偶，更多的時候抬頭看她做飯。

我曾經對外婆說，我長大後要像您啊！
能夠擁有自己的工作間，很幸運呢！

是因為屋子小，又或立志要像外婆？
我的眼睛總離不開外婆的一雙手，
口裡更喜歡問外婆這個料理、那個作法。
外婆做菜粿嗎？我也趕著做。
外婆做甜湯嗎？我也擠進廚房去。

沒有十足的手藝，卻有八分的相似。
外婆，您知道嗎？
我真的做到了，做到一個很像您的我。

Chapter 5
蔬菜料理

不只是肉類的烹調，也要補足對蔬菜的營養需求，餃子、
煎餅、春卷、天婦羅……蔬菜也能做出充滿變化的好滋味！

晚餐　便當

日式韭菜雞肉餃子

韭菜油揚鍋

兩餐前準備 Preparation

材料
韭菜 150 克
豆芽菜 100 克

Ingredients
Chive 150g
Bean Sprouts 100g

 作法 Procedures

1. 韭菜洗淨擦乾。
2. 豆芽菜去頭及根部，洗淨。
1.Rinse and drain chive.
2.Remove root from bean sprouts, rinse.

韮菜油揚鍋
Chive and Deep Fried Tofu in Pot

日式韮菜
雞肉餃子
Japanese Chive and Chicken Dumplings

晚餐

韭菜油揚鍋
Chive and Deep Fried Tofu in Pot

材料
已準備的韭菜 75 克
已準備的豆芽菜 50 克
油揚豆腐皮 2 塊
紅蘿蔔 1/2 條

Ingredients
Prepared Chive 75g
Prepared Bean Sprouts 50g
Japanese Deep Fried Tofu 2 pcs
Carrot 1/2 pcs

醬汁
日式醬油 1 湯匙
味醂 1 湯匙
料理酒 1 湯匙
水 100 毫升
鰹魚粉 1 茶匙

Sauce
Japanese Soy Sauce 1 tbsp
Mirin 1 tbsp
Sake 1 tbsp
Water 100ml
Dashi Powder 1 tsp

作法
1. 油揚豆腐皮整塊放小鍋內汆燙，盛起，切成細條。
2. 韭菜切成 5 公分長的小段；紅蘿蔔切成 5 公分長細條。
3. 小鍋裡加所有醬汁煮滾，放韭菜、豆芽菜、油揚豆腐皮及紅蘿蔔絲同煮約 10 分鐘。

Procedures
1.Rinse Japanese deep fried tofu and scald in boiling water for a short while. Rinse again. Thinly slice.
2.Cut chive into 5cm long. Cut carrot into 5cm long shreds.
3.Add seasoning in a small pot, bring to boil. Add chive, bean sprouts, tofu and carrot, simmer for 10 minutes.

  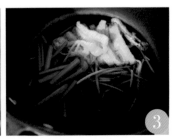

 料理小提示 Cooking Tips

油揚豆腐皮在一般日式超市有販售，也可以用我們的油豆腐代替，同樣切成細條及汆燙備用。
Japanese deep fried tofu can be bought in Japanese supermarket, or it can be replaced by Chinese deep fried tofu.

簡易的日式素菜鍋料理，說是鍋，其實是一道蔬菜煮切，多作為前菜或配菜。

很適合當作便當主食，又或假日午餐。
可以大量製作，放到冰箱儲存，有空時
煎煮來吃，比超級市場的好吃多倍。

日式韭菜雞肉餃子
Japanese Chive and Chicken Dumplings

材料
已準備的韭菜 75 克
已準備的豆芽菜 50 克
雞絞肉 100 克
冬菇 2 朵
蔥 1 株
餃子皮 15 片

調味料
日式醬油 2 湯匙
味醂 1 湯匙
料理酒 1 湯匙
糖 2 茶匙
玉米粉 1 湯匙

Ingredients
Prepared Chive 75g
Prepared Bean Sprouts 50g
Minced Chicken 100g
Shitake mushroom 2 pcs
Green Onion 1 stalk
Dumpling wrappers 15 pcs

Seasoning
Japanese Soy Sauce 2 tbsp
Mirin 1 tbsp
Sake 1 tbsp
Sugar 2 tsp
Corn Starch 1 tbsp

作法
1. 冬菇以熱水泡發，泡軟後切小丁。
2. 韭菜、豆芽菜及蔥切碎。
3. 大碗內，放入所有材料（餃子皮除外）及調味料拌勻成餡料，並攪拌至帶有黏性。
4. 在餃子皮中央放上作法 3 的餡料，餃子皮邊緣沾水，對摺。
5. 平底鍋內加少許油，放進餃子略煎，倒入水半杯，蓋上，轉小火煎煮至水乾。（餃子必須加蓋半煎煮才可以熟透。）
6. 開蓋，轉大火煎至餃子底部呈金黃色。

Procedures
1. Rinse mushroom and soak toll tender. Squeeze dry. Dice.
2. Finely chop chive, bean sprouts and green onion.
3. Mix all ingredients（except dumpling wrapper）and marinade in a big bowl.
4. Place a spoonful of meat paste in the middle of dumpling wrapper, coat edge with water, fold the dumpling wrapper into half.
5. Heat frying pan with oil, pan fry dumplings. Pour in half cup of water. Cover and simmer until water dried out.（Dumplings can only be cooked by simmer with water.）
6. Remove cover, turn to high heat till dumpling golden brown.

 料理小提示 Cooking Tips

便當裡的餃子可以煎，也可以炸，隨自己喜歡的口味吧！
Dumplings can be pan fried or deep fried, according to your own preference.

晚餐 韓風南瓜炒豬肉

便當 韓風南瓜煎餅

兩餐前準備 Preparation

材料
日本南瓜 1/2 個

Ingredients
Japanese Pumpkin 1/2 pcs

醃料
鹽 2 茶匙

Marinade
Salt 2 tsp

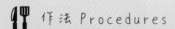 作法 Procedures

1. 南瓜去籽。
2. 將南瓜切成薄片。
3. 其中一半的南瓜片切成細絲,加鹽略醃至南瓜變軟出水。

1. Remove seeds from pumpkin.
2. Thinly slice pumpkin.
3. Shred half of portion of pumpkin, marinade with salt.

 韓風南瓜
炒豬肉
*Korean Style Stir-fry Pork
Tenderloin with Pumpkin*

韓風南瓜煎餅
*Korean Style Pumpkin
Pancake*

晚餐

韓風南瓜炒豬肉
Korean Style Stir-fry Pork Tenderloin with Pumpkin

材料
已準備的切片南瓜
薄切嫩肩豬肉片 100 克
泡菜 100 克
蒜蓉 1 茶匙

調味料
日式醬油 1 湯匙
糖 2 茶匙

Ingredients
Prepared Thinly Sliced Pumpkin Thinly
Sliced Pork Tenderloin 100g
Korean Kimchi 100g
Minced Garlic 1 tsp

Seasoning
Japanese Soy Sauce 1 tbsp
Sugar 2 tsp

作法
1. 平底鍋放油燒熱，爆香蒜蓉。
2. 放嫩肩豬肉片及調味料炒至肉全熟，盛起。
3. 在同一個平底鍋內放南瓜，加少許鹽煎至兩面金黃。
   （先將南瓜煎至金黃，可以保持香脆質感。）
4. 加入泡菜略炒，肉片回鍋，大火炒勻。

Procedures
1. Heat frying pan with oil. Saute garlic.
2. Stir fry pork tenderloin slice and seasoning till cooked, set aside.
3. Pan fry pumpkin slice in same frying pan, drizzle with salt. Cook till golden
   brown.（Brown pumpkin helps keeping the crispy texture.）
4. Add kimchi and pork tenderloin slice, stir fry over high heat.

 料理小提示 Cooking Tips

若希望韓國風味再重一點，可以加少許麻油炒勻。
Add a droplet of sesame oil to enrich the Korean favor.

南瓜的鮮甜，加上泡菜的香辣，這是一
道滋味很豐富的料理，簡單以一個蔬菜
湯搭配，就成為簡單美味的晚餐。

149

為便當加點新意，將主食由白飯變作
韓式煎餅，給他一個干飯小驚喜。

韓風南瓜煎餅
Korean Style Pumpkin Pancake

材料
已準備的南瓜絲
泡菜 50 克
雞蛋 1 顆
麵粉 100 克
水 40 毫升

Ingredients
Prepared Shredded Pumpkin
Kimchi 50g
Egg 1 pc
Flour 100g
Water 40ml

調味料
鹽 1 茶匙
味醂 2 茶匙
糖 1 茶匙

Seasoning
Salt 1tsp
Mirin 2 tsp
Sugar 1 tsp

作法
1. 雞蛋、麵粉及水放碗內拌勻。
2. 拌入已泡軟的南瓜絲、泡菜及調味料，拌勻成麵糊。
   （如麵糊太稠，可以自行加水調節。）
3. 平底鍋加油燒熱，倒入麵糊，將麵糊推開，盡量令南瓜餅變薄。
4. 煎 2 ～ 3 分鐘至金黃，翻面再將另一面煎至金黃。
5. 放涼後切成適合大小。

Procedures
1.Mix egg, flour and water in a bowl.
2.Add shredded pumpkin, kimchi and seasoning, stir well.
   （Add water if the batter is too dry.）
3.Heat frying pan with oil, pour in batter. Try to spread batter all over frying pan.
4.Pan fry for 2-3 minutes till golden brown. Turn over to another side.
5.Cut after cooled down.

 料理小提示 Cooking Tips
切成小正方狀，感覺更韓式，放進便當盒內也更容易。
For a better Korean presentation, cut pancake into small square.

晚餐 茄子雜炒

便當 香麻扒茄子

兩餐前準備 Preparation

材料
茄子 4 條

Ingredients
Eggplant 4 pcs

作法 Procedures

茄子洗淨，斜切成塊，再切成粗條。
Rinse eggplant, slice in diagonal way, shred.

🥢 茄子雜炒
Stir Fry Eggplant with
Assorted Vegetables

 香麻扒茄子
Eggplant Steak in
Sesame Favor

晚餐

茄子雜炒

Stir Fry Eggplant with Assorted Vegetables

材料
已準備的茄子 2 個
番茄 1 個
杏鮑菇 2 個
四季豆 2 根
蒜蓉 1 茶匙

Ingredients
Prepared Eggplant 2 pcs
Tomato 1 pc
Eryngii Mushroom 2 pcs
Green Beans 4 pcs
Minced Garlic 1 tsp

調味料
日式醬油 1 湯匙
味醂 1 湯匙
料理酒 1 湯匙
糖 1 茶匙

Seasoning
Japanese Soy Sauce 1 tbsp
Mirin 1 tbsp
Sake 1 tbsp
Sugar 1 tsp

作法
1. 番茄切成 8 等份；杏鮑菇切厚片；四季豆切小段。
2. 平底鍋放油燒熱，爆香蒜蓉。
3. 加入茄子、杏鮑菇及四季豆炒至軟。
4. 放進番茄及調味料大炒至收汁。

Procedures
1.Cut Tomato into 8 portions. Thickly slice eringi mushroom. Cut green bean into small section.
2.Heat oil in frying pan. Saute garlic.
3.Stir fry eggplant, eringi mushroom and green bean till tender.
4.Add tomato and seasoning. Stir fry over high heat till sauce thicken.

 料理小提示 Cooking Tips

若喜歡吃肉，可以添加少許牛絞肉一同烹煮。
Cook with ground beef if you are a meat lover.

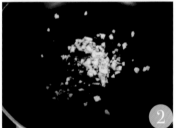

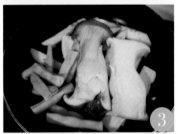

不同口感的蔬菜放在一起，是什麼滋味？
這道料理給你一個完美的答案。

把茄子裹上麵包粉及芝麻炸得
香脆，做出不一樣的口感。

香麻扒茄子

Eggplant Steak in Sesame Favor

材料
已準備的茄子 2 個
雞蛋 1 顆
麵包粉 1/2 碗
白芝麻 2 湯匙

調味料
鹽 2 茶匙
胡椒粉 1 茶匙
玉米粉 1 湯匙

Ingredients
Prepared Eggplant 2 pcs
Egg 1 pc
Bread Crumbs 1/2 bowl
Sesame 2 tbsp

Seasoning
Salt 2 tsp
Pepper 1 tsp
Corn Starch 1 tbsp

作法
1. 打發雞蛋；麵包粉及白芝麻拌勻。
2. 大碗內放茄子,加調味料拌勻。
3. 茄子依序沾上蛋液及預先混勻的芝麻麵包粉。
4. 小鍋內加油燒滾,放進茄子炸至金黃。

Procedures
1. Beat egg. Mixed bread crumbs and sesame.
2. Add eggplant and marinade in a bowl. Mix well.
3. Coat eggplant with beaten egg and bread crumbs.
4. Boil oil in small pot, deep fry eggplant till golden brown.

 料理小提示 Cooking Tips

茄子要選又肥又短的,會較硬挺,更適合煎炸。
Eggplant of short and fat would be more suitable for deep fried.

晚餐 便當

苦瓜咖哩

天婦羅

和風

苦瓜炒蛋

兩餐前準備 Preparation

材料
苦瓜 2 個

Ingredients
Bitter Gourd 2 pcs

🍴 作法 Procedures

1. 苦瓜洗淨，切半去籽。
2. 其中一條苦瓜切成薄片，另一條切厚片。

1.Rinse bitter gourd, cut into half. Remove seed.
2.Thinly slice one of the bitter gourd, thickly slice another one.

和風苦瓜炒蛋
Japanese Style Stir Fry
Bitter Gourd with Egg

苦瓜咖哩天婦羅
Bitter Gourd Curry Tempura

晚餐

和風苦瓜炒蛋
Japanese Style Stir Fry Bitter Gourd with Egg

材料
已準備的苦瓜薄片
薄切嫩肩豬肉片 100 克
紅蘿蔔 1/2 條
雞蛋 2 顆

Ingredients
Prepared Thinly Sliced Bitter Gourd
Thinly Sliced Pork Tenderloin 100g
Carrot 1/2 pcs
Egg 2 pcs

醃料
日式醬油 1 湯匙
味醂 1 湯匙
糖 1 茶匙

Marinade
Japanese Soy Sauce 1 tbsp
Mirin 1 tbsp
Sugar 1 tsp

調味料
味醂 1 湯匙
鹽 1 茶匙

Seasoning
Mirin 1 tbsp
Salt 1 tbsp

作法
1. 紅蘿蔔切絲；雞蛋在小碗內打發；嫩肩豬肉片加醃料醃至少 20 分鐘。
2. 平底鍋放油燒熱，加肉片炒熟，盛起。
3. 同一個平底鍋內放進苦瓜及紅蘿蔔同炒。
4. 加蛋液、肉片回鍋，加調味料大火快炒。
   （雞蛋必須大火快炒，才能夠做到散開的效果。）

Procedures
1. Thinly slice carrot. Beat eggs in a small bowl. Marinate pork tenderloin slice for 20 minutes.
2. Heat oil in frying pan. Add pork tenderloin slice, stir fry till cooked, set aside.
3. Stir fry bitter gourd and carrot in same frying pan.
4. Add beaten eggs and pork tenderloin slice. Add seasoning and stir fry still sauce thicken.（Egg must be stirred fry over high heat to shorten the cooking time.）

 料理小提示 Cooking Tips
苦瓜用鹽醃過之後，可減輕苦澀味。
Marinade bitter gourd with salt will reduce bitterness.

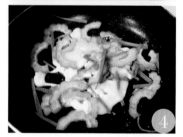

苦瓜炒蛋在香港家庭很常見，
試一下這道和風苦瓜炒蛋，感
受相同材料變出的不同風味。

161

天婦羅是很常見的一道料理，但用上苦瓜，就變得獨特了。加些許咖哩粉，調和一下苦味，讓苦瓜變得更容易親近。

苦瓜咖哩天婦羅
Bitter Gourd Curry Tempura

材料
已準備的苦瓜厚片

Ingredients
Thickly slice Bitter Gourd

天婦羅麵糊
天婦羅粉 4 湯匙
咖哩粉 2 茶匙
水 6 湯匙
鹽 1 茶匙

Seasoning
Tempura flour 4 tbsp
Japanese Curry Powder 2 tsp
Water 6 tbsp
Salt 1 tsp

作法
1. 大碗內將天婦羅麵糊混合。
2. 小鍋內加油燒熱。
3. 將苦瓜逐一沾上混勻的天婦羅麵糊。
4. 放入滾油內炸至金黃。（天婦羅必須以大火快炸才能保持酥脆口感。）

Procedures
1. Mix tempura batter in a bowl.
2. Boil oil in a small pot.
3. Coat bitter gourd with tempura batter.
4. Deep fry till golden brown.
（Tempura must be deep fried over high heat to keep the crispy texture.）

 料理小提示 Cooking Tips

更喜愛原味天婦羅嗎？不加咖哩粉就可以了。
Fancy for original tempura, simply omit curry powder when mixing batter.

晚餐　便當

和風炒野菜　素春卷

兩餐前準備 Preparation

材料
高麗菜 1/2 個
洋蔥 1/2 個
紅蘿蔔 1 條
金針菇 1 包

Ingredients
Cabbage 1/2 pcs
Onion 1/2 pcs
Carrot 1 pc
Enokitake mushroom 1 pack

 作法 Procedures

1. 將所有蔬菜洗淨。
2. 高麗菜及洋蔥切絲；紅蘿蔔切細絲；金針菇去蒂頭尾部。
1.Rinse and drain all vegetables.
2.Thinly slice cabbage and onion. Shred carrot. Remove root of enokitake mushroom.

 和風炒野菜
Japanese Style Stir Fry
Assorted Vegetables

素春卷
Vegetarian Spring Roll

晚餐

和風炒野菜

Japanese Style Stir Fry Assorted Vegetables

材料
已準備的蔬菜（全部）
白芝麻 1 茶匙
七味粉 1 茶匙

Ingredients
Prepared Assorted Vegetables （Whole Portion）
Sesame 1 tsp
Japanese Chili Powder 1 tsp

調味料
味醂 1 湯匙
鹽 1 茶匙
糖 1 茶匙

Seasoning
Mirin 1 tbsp
Salt 1 tsp
Sugar 1 tsp

作法
1. 平底鍋放油燒熱，炒洋葱至軟。
2. 下紅蘿蔔略炒，再加入高麗菜及金針菇。
3. 加入調味料，大火炒至所有蔬菜變軟及全熟。
4. 吃時撒上白芝麻及七味粉。（記得要留下一半的蔬菜用作翌日的便當。）

Procedures
1.Heat frying pan with oil. Stir fry onion till tender.
2.Add carrot, stir fry. Add cabbage and enokitake mushroom.
3.Add seasoning, stir fry vegetable till tender.
4.Serve with sesame and Japanese chili powder. （Remember to keep half of the cooked vegetables for lunchbox.）

 料理小提示 Cooking Tips

這道炒野菜很隨性，除了高麗菜是必需外，其他的配菜可根據自己的口味更換。
Beside cabbage, you may change to any kind of vegetables according to your preference.

用途很廣的一道蔬菜料理，可以單吃當作主菜，也可以跟拉麵或烏龍麵同炒。

我們愛把春卷當小吃，但日本人卻愛
把它當主食。以蔬食入菜的春卷，少
了油膩，多了清新。

 素春卷
Vegetarian Spring Roll

材料
「和風炒野菜」預留的蔬菜
小春卷皮 10 張

Ingredients
Reserved Vegetable from Previous Recipe
Spring Roll Pastry 10 pcs

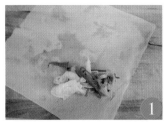

作法
1. 春卷皮先放室溫半小時解凍，再慢慢撕開。
2. 春卷皮擺成菱形平放，將一小份蔬菜放在底端，再捲起春卷皮直至包裹蓋住整份蔬菜。
3. 兩邊的春卷皮向內摺後，再繼續向上捲起，在開口沾點水黏住接合處。
4. 平底鍋加油燒熱，放入春卷以小火炸至金黃。
   （火太猛，春卷會很容易變焦。）

Procedures
1. Defrost spring roll pastry in room temperature for around half hour.
2. Place spring roll pastry in diamond shape. Add a portion of vegetable at the corner. Roll up to half.
3. Fold two side pastry to center. Keep rolling up and glue the end with water.
4. Deep fry spring roll in boiling oil till golden brown.
   （Spring roll will be easily burnt over high heat.）

 料理小提示 Cooking Tips
待春卷完全放涼後再蓋上便當盒，就可以保持春卷香脆。
Put spring roll inside lunchbox after it completely cool.

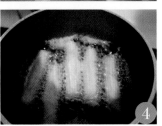

Chapter 6
常備菜

就現有食材，隨手製成簡單易保存的常備菜，搭配主菜隨即擁有一頓豐盛饗宴。

常備菜，就是經常配備在冰箱或家裡的菜式，
由隨手取來的食材做成。
不僅作法簡單，也容易保存。

它可以當作晚飯小菜，也可以放進便當盒內。
沒有豐富的味道，卻有點綴主菜的力量，
把料理變得簡單輕鬆。

常備菜

雜錦蛋捲

Rolled Omelet with Assorted Vegetables

材料
雞蛋 2 顆
煮過的菜葉或其他現有的已熟食材

調味料
日式醬油 2 茶匙
味醂 1 茶匙

Ingredients
Egg 2 pcs
Cooked vegetable leaves and other cooked food

Seasoning
Japanese Soy Sauce 1 tsp
Mirin 1 tsp

作法
1. 雞蛋加調味料打發。
2. 在日式的長形煎鍋內加少許油，稍微燒熱，倒入 1/3 的蛋汁。
3. 蛋汁凝固後，在底端放上菜葉或其他材料，用筷子捲起蛋皮，壓實。
4. 再倒入另外的 1/3 蛋液，待稍微凝固後，捲起蛋卷；重複完成剩下的 1/3 蛋液。
5. 放涼後切塊。

Procedures
1. Beat eggs with marinade.
2. Heat a rectangular frying pan with oil, pour in about 1/3 of eggs.
3. When egg almost done, place vegetables leaves or other cooked food at one end, roll up by chopsticks, press tight.
4. Pour another 1/3 of egg, roll up when done. Repeat the last layer.
5. Slice after cool down.

常備菜

薑蒜炒蝦仁

Fried Prawns with Ginger and Garlic

材料
雪藏蝦仁 10 隻
薑 2 片
蒜蓉 1 茶匙

調味料
日式醬油 1 茶匙
味醂 1 茶匙
料理酒 1 茶匙
糖 1 茶匙

Ingredients
Frozen Prawn 10 pcs
Ginger 2 slices
Garlic 1 tsp

Seasoning
Japanese Soy Sauce 1 tsp
Mirin 1 tsp
Sake 1 tsp
Sugar 1 tsp

作法
1. 雪藏蝦仁解凍後洗淨；薑切成碎末。
2. 平底鍋放油燒熱，爆香薑末及蒜蓉。
3. 放進蝦仁及調味料一起拌炒至收汁。

Procedures
1.Chill and drain prawns. Dice ginger.

2.Heat frying pan with oil, saute ginger and garlic.

3.Add prawns and marinade, stir fry till sauce thicken.

常備菜

汁燒京葱竹輪卷

Teriyaki Leek with Chikuwa

材料
迷你竹輪卷 6 條
京葱（大葱）1 株

調味料
日式醬油 2 茶匙
味醂 2 茶匙
糖 1 茶匙

Ingredients
Mini Chikuwa 6 pcs
Leek 1 pc

Seasoning
Japanese Soy Sauce 2 tsp
Mirin 2 tsp
Sugar 1 tsp

作法
1. 迷你竹輪直切對半，京葱切段，跟迷你竹輪約相同長度。
2. 平底鍋放油燒熱，加入竹輪及京葱同炒。
3. 加調味料大火炒至收汁。

Procedures
1.Cut chikuwa in half, cut leek into same length as chikuwa.
2.Heat frying pan with oil, stir fry chikuwa and leek.
3.Add marinade and cook till the sauce thicken.

培根秋葵卷

Okra roll with Bacon

材料
培根 3 片
秋葵 6 根

Ingredients
Bacon 3 pcs
Okra 6 pcs

作法
1. 秋葵去頭部，以刀背刮去表面的細毛。
2. 培根切半，利用半片培根捲起秋葵 1 根，其餘重複同樣動作。
3. 平底鍋放油燒熱，將培根秋葵卷的開口朝下，煎至培根全熟。

Procedures
1.Slice off stem from okra. Remove hair of okra by knife.
2.Cut bacon into half. Roll okra by half piece of bacon. Repeat the same procedure.
3.Heat frying pan with oil, place the opened end of bacon roll down. Cook till done.

常備菜

奶油炒蔬菜

Stir Fry Veggie with Butter

材料
紅蘿蔔 1/2 條
四季豆 2～3 根
花椰菜 1/2 個
奶油 1 湯匙

調味料
鹽 1 茶匙
糖 2 茶匙

Ingredients
Carrot 1/2 pcs
Green Bean 2-3 pcs
Cauliflower 1/2 pcs
Butter 1 tbsp

Seasoning
Salt 1 tsp
Sugar 2 tsp

作法
1. 紅蘿蔔去皮切約 1 公分厚片;四季豆切段,約 3 公分長;花椰菜切成小顆。
2. 全部蔬菜放進滾水煮熟,瀝乾。
3. 小鍋內放進奶油煮至融化,加入蔬菜及調味料大炒。

Procedures
1. Slice carrot into 1cm thick. Green bean cut into 3 cm pieces.

   Cut cauliflower into small pieces.
2. Bring all vegetables to boil. Drain well.
3. Melt butter in a small pot, stir fry vegetable with marinade.

味噌青花菜

Broccoli with Miso

材料
青花菜 1 個

調味料
味噌 1 湯匙
糖 2 茶匙
水 1 湯匙

Ingredients
Broccoli 1 pc

Seasoning
Miso 1 tbsp
Sugar 2 tsp
Water 1 tbsp

作法
1. 青花菜切小顆，洗淨，以滾水煮熟，瀝乾水分。
2. 小鍋內略燒煮調味料，放入青花菜拌勻。

Procedures
1.Cut broccoli into small pieces. Rinse and boil, drain well.
2.Bring marinade to boil, add broccoli, stir well.

{ 品味
  生活 }

200 道美味湯品輕鬆做：
幸福濃湯╳健康高湯╳各國好湯
莎拉‧陸慧絲 著／諾然 譯／定價 350 元

清新爽口的西芹蘋果湯， 惹味香濃的椰汁咖哩雞湯，
層次豐富的酪梨酸奶油湯， 應節佳品聖誕雜菜湯， 健
康有益的扇貝花椰菜湯……等， 隨著本書，任何場合、
任何季節， 都能嘗到道地異國風味的好湯！

200 道健康咖哩輕鬆做：
濃郁湯品╳辛香料沙拉╳開胃小點╳美味主食
蘇尼爾 ‧ 維查耶納伽爾 著／陳愛麗 譯／定價 350 元

清爽的香料馬鈴薯蘋果沙拉，口感濃郁的薄荷菠菜酸
奶湯，極具特色的泰式叢林咖哩鴨，大飽口福的香料
扁豆印度咖哩飯……等，來自世界各地的風味，讓新
手也能跟著食譜烹調出異國美味！

200 道義大利麵料理輕鬆做：
溫暖湯品╳簡易沙拉╳美味麵食
瑪莉雅 ‧ 芮奇 著／謝映如 譯／定價 350 元

簡單的通心粉沙拉，經典的肉醬義大利千層麵，奢侈
的朝鮮薊心麵……等，本書收錄海鮮、畜肉、禽肉、
素食義大利麵主食，到一般台灣人少做的義大利麵湯
與沙拉，除了傳統作法，還有變化、創新，滿足每個
喜愛義大利麵的朋友。

200 道鍋煮美食輕鬆做：
惹味肉香鍋╳香濃海鮮鍋╳各國好湯
喬安娜‧費羅 著／關仰山 譯／定價 350 元

別具特色的地中海豬肉砂鍋，香濃惹味的墨西哥醬雞，
鮮甜又吸睛的墨魚黑米飯，以及健康有益的西西里燉
茄子……等，一個鍋煮好一頓飯，惹味鮮肉、特色野
味、香濃海鮮、健康素菜，四種不同主題，切合你不
同需要！

異國料理

印度料理初學者的第一本書：
印度籍主廚奈爾善己教你做 70 道印度家常料理
奈爾善己　著／陳柏瑤　譯／定價 320 元

日本超人氣印度料理老店，傳家食譜不藏私公開！連印度人都說超．好．吃～從南北咖哩、配菜到米飯麵包、甜點……怎麼切、怎麼炒，文字步驟配詳盡照片，掌握基本 3 步驟，新手也能做出印度本格菜！

異國風主食料理：焗烤、燉飯、粥品、鍋物等
60 道美味幸福上桌（中英對照）
洪白陽（CC 老師）　著／楊志雄　攝影／定價 385 元

歐式義大利麵、燉飯，東南亞風味河粉湯、海南雞飯，日韓料理築地海鮮粥、高升元寶湯，中港台料理荷香糯米飯、客家粿條蒸鮮蝦等，全書 60 道主食、10 款基本醬汁與高湯，以中英對照詳細介紹最受歡迎的異國主食料理，隨著CC老師，美味，幸福上桌！

Paco 上菜：
西班牙美味家常料理
Mr. Paco　著／蕭維剛　攝影／定價 340 元

地中海料理素來是長壽和健康的飲食代表；其中便包含西班牙料理，運用大量橄欖油、大蒜、瓜果、甜椒、海鮮、穀類植物和適量的葡萄酒烹調。本書以擁有西班牙血統的 Paco 主廚掌杓，示範最具在地風情的家常料理，分為開胃菜、主食、湯品、甜點、飲料，共 40 多道，讓你在家也能品嘗西班牙家常美味。

巴黎日常料理：
法國媽媽的美味私房菜 48 道
殿真理子　著／程馨頤　譯／定價 300 元

和你分享 48 道法國正統家庭料理╳你不知道的法國餐桌二三事！法式鹹可麗餅、甜蜜杏桃塔、櫻桃克拉芙緹、普羅旺斯燉鮮蔬……以及最天然的季節果醬祕方、釀鮮蔬撇步，輕鬆上手，簡單易做，從餐前菜到甜點，享受專屬於法式的慢食美味。

地址： 　　　縣/市　　　　鄉/鎮/市/區　　　　路/街

　　　段　　　巷　　　弄　　　號　　　樓

廣　告　回　函
台北郵局登記證
台北廣字第2780號

三友圖書有限公司　收
SANYAU PUBLISHING CO., LTD.

106　　台北市安和路2段213號4樓

三友圖書
讀書俱樂部

購買《晚餐與便當一次搞定：1次煮2餐的日式常備菜》的讀者有福啦！只要詳細填寫背面問卷，並寄回三友圖書，即有機會獲得 COOK-ZEN 微波專用魔法鍋！

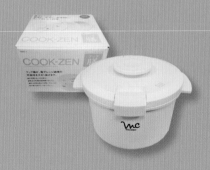

COOK-ZEN 微波專用魔法鍋 乙個
價值 2,000 元（共兩名）

活動期限至 2016 年 1 月 18 日為止，
詳情請見回函內容。

＊本回函影印無效

四塊玉文創╳橘子文化╳旗林文化╳食為天文創
https://www.facebook.com/comehomelife
http://www.ju-zi.com.tw

親愛的讀者：

感謝您購買《晚餐與便當一次搞定：1 次煮 2 餐的日式常備菜》一書，為回饋您對本書的支持與愛護，只要填妥本回函，並於 2016 年 1 月 18 日寄回本社（以郵戳為憑），即有機會抽中「COOK-ZEN 微波專用魔法鍋」乙個（共兩名）。

姓名 _____ 出生年月日_____

電話 _____ E-mail _____

通訊地址_____

臉書帳號 _____

部落格名稱 _____

1 年齡
□ 18 歲以下 □ 19 歲～ 25 歲 □ 26 歲～ 35 歲 □ 36 歲～ 45 歲 □ 46 歲～ 55 歲
□ 56 歲～ 65 歲 □ 66 歲～ 75 歲 □ 76 歲～ 85 歲 □ 86 歲以上

2 職業
□軍公教 □工 □商 □自由業 □服務業 □農林漁牧業 □家管 □學生
□其他 _____

3 您從何處購得本書？
□網路書店 □博客來 □金石堂 □讀冊 □誠品 □其他 _____
□實體書店 _____

4 您從何處得知本書？
□網路書店 □博客來 □金石堂 □讀冊 □誠品 □其他 _____
□實體書店 _____ □ FB(微胖男女粉絲團 - 三友圖書)
□三友圖書電子報 □好好刊（季刊） □朋友推薦 □廣播媒體 _____

5 您購買本書的因素有哪些？（可複選）
□作者 □內容 □圖片 □版面編排 □其他 _____

6 您覺得本書的封面設計如何？
□非常滿意 □滿意 □普通 □很差 □其他 _____

7 非常感謝您購買此書，您還對哪些主題有興趣？（可複選）
□中西食譜 □點心烘焙 □飲品類 □旅遊 □養生保健 □瘦身美妝 □手作 □寵物
□商業理財 □心靈療癒 □小說 □其他 _____

8 您每個月的購書預算為多少金額？
□ 1,000 元以下 □ 1,001 ～ 2,000 元 □ 2,001 ～ 3,000 元 □ 3,001 ～ 4,000 元
□ 4,001 ～ 5,000 元 □ 5,001 元以上

9 若出版的書籍搭配贈品活動，您比較喜歡哪一類型的贈品？（可選 2 種）
□食品調味類 □鍋具類 □家電用品類 □書籍類 □生活用品類 □ DIY 手作類
□交通票券類 □展演活動票券類 □其他 _____

10 您認為本書尚需改進之處？以及對我們的意見？
_____

感謝您的填寫，
您寶貴的建議是我們進步的動力！

◎本回函得獎名單公布相關資訊
得獎名單抽出日期：2016 年 1 月 21 日
得獎名單公布於：
・臉書「微胖男女編輯社 - 三友圖書」https://www.facebook.com/comehomelife
・痞客邦「微胖男女編輯社 - 三友圖書」http://sanyau888.pixnet.net/blog